AF880300

ENERGY SCIENCE, ENGINEERING AND TECHNOLOGY

DAMS, DROUGHT AND ENERGY-WATER INTERDEPENDENCIES

ENERGY SCIENCE, ENGINEERING AND TECHNOLOGY

Additional books in this series can be found on Nova's website under the Series tab.

Additional E-books in this series can be found on Nova's website under the E-book tab.

ENERGY SCIENCE, ENGINEERING AND TECHNOLOGY

DAMS, DROUGHT AND ENERGY-WATER INTERDEPENDENCIES

DANIEL L. CALZI
EDITOR

nova publishers
New York

Copyright © 2013 by Nova Science Publishers, Inc.

All rights reserved. No part of this book may be reproduced, stored in a retrieval system or transmitted in any form or by any means: electronic, electrostatic, magnetic, tape, mechanical photocopying, recording or otherwise without the written permission of the Publisher.

For permission to use material from this book please contact us:
Telephone 631-231-7269; Fax 631-231-8175
Web Site: http://www.novapublishers.com

NOTICE TO THE READER

The Publisher has taken reasonable care in the preparation of this book, but makes no expressed or implied warranty of any kind and assumes no responsibility for any errors or omissions. No liability is assumed for incidental or consequential damages in connection with or arising out of information contained in this book. The Publisher shall not be liable for any special, consequential, or exemplary damages resulting, in whole or in part, from the readers' use of, or reliance upon, this material. Any parts of this book based on government reports are so indicated and copyright is claimed for those parts to the extent applicable to compilations of such works.

Independent verification should be sought for any data, advice or recommendations contained in this book. In addition, no responsibility is assumed by the publisher for any injury and/or damage to persons or property arising from any methods, products, instructions, ideas or otherwise contained in this publication.

This publication is designed to provide accurate and authoritative information with regard to the subject matter covered herein. It is sold with the clear understanding that the Publisher is not engaged in rendering legal or any other professional services. If legal or any other expert assistance is required, the services of a competent person should be sought. FROM A DECLARATION OF PARTICIPANTS JOINTLY ADOPTED BY A COMMITTEE OF THE AMERICAN BAR ASSOCIATION AND A COMMITTEE OF PUBLISHERS.

Additional color graphics may be available in the e-book version of this book.

Library of Congress Cataloging-in-Publication Data

ISBN: 978-1-62808-942-4

Published by Nova Science Publishers, Inc. † New York

CONTENTS

PREFACE

Chapter 1 – The U.S. Department of Energy (DOE) and the U.S. Department of Homeland Security (DHS) collaborated to examine the interdependencies between two critical infrastructure sectors – Dams and Energy. The study highlights the importance of hydroelectric power generation, with a particular emphasis on the variability of weather patterns and competing demands for water which determine the water available for hydropower production. In recent years, various regions of the Nation suffered drought, impacting stakeholders in both the Dams and Energy Sectors. Droughts have the potential to affect the operation of dams and reduce hydropower production, which can result in higher electricity costs to utilities and customers. Conversely, too much water can further complicate the operation of dams in ways that can be detrimental to hydropower production and to the infrastructure of the dams. Discussions with dam owners and operators revealed that the storage capacity and conveyance flexibility of most conventional hydroelectric facilities were designed to accommodate local or regional historical patterns of hydrologic variability. Thus, episodic low water conditions, as opposed to long-term drought conditions, are not critical contributors to reduced hydropower production; however, the requirements for providing sufficient water for irrigation, environmental protection, transportation, as well as community and industrial uses are already in conflict in many places. Low water conditions (e.g., drought) and high water conditions (e.g., flood) resulting from extreme weather variability can strain the operation of dams and heighten the degree of competition for available water. Although hydroelectric facilities are a type of asset that falls under the auspices of the Dams Sector, they are also an important element to the Energy Sector because the electric power they generate is critical to maintaining the reliability of the Nation‘s electricity supply. Therefore, this joint effort underscores the value of a cross-sector partnership model in the identification and discussion of issues significant to dam and utilities owners and operators, through which can help enhance their resilience against the potential impacts associated with the variability of weather patterns and extreme fluctuations of water flow.

Chapter 2 – Report on National Energy Technology Laboratory publication, on An Analysis of the Effects of Drought Conditions on Electric Power Generation in the Western United States, dated April 2009.

Chapter 3 – Tensions between the energy and water sectors occur when demand for electric power is high and water supply levels are low. There are several regions of the country, such as the western and southwestern states, where the confluence of energy and water is always strained due to population growth. However, for much of the country, this tension occurs at particular times of year (e.g., summer) or when a region is suffering from drought conditions. This report discusses prior work on the interdependencies between energy and water. It identifies the types of power plants that are most likely to be susceptible to water shortages, the regions of the country where this is most likely to occur, and policy options that can be applied in both the energy and water sectors to address the issue. The policy options are designed to be applied in the near term, applicable to all areas of the country, and to ease the tension between the energy and water sectors by addressing peak power demand or decreased water supply.

Chapter 4 – Drought is a natural hazard with often significant societal, economic, and environmental consequences. Public policy issues related to drought range from how to identify and measure drought to how best to prepare for, mitigate, and respond to drought impacts, and who should bear associated costs. Severe drought in 2011 and 2012 fueled congressional interest in near-term issues, such as current (and recently expired) federal programs and their funding, and long-term issues, such as drought forecasting and various federal drought relief and mitigation actions. Continuing drought conditions throughout the country contribute to ongoing interest in federal drought policies and responses. As of April 2013, drought has persisted across approximately two-thirds of the United States and is threatening agricultural production and other sectors. More than 1,180 counties so far have been designated as disaster areas for the 2013 crop season, including 286 counties contiguous to primary drought counties. In comparison, in August 2012, more than 1,400 counties in 33 states had been designated as disaster counties by the U.S. Secretary of Agriculture. Most attention in the 112th Congress focused on the extension of expired disaster assistance programs in separate versions of a 2012 farm bill. Attention in the 113th Congress again is expected to focus on farm bill legislation; however, other bills addressing different aspects of drought policy and response have also been introduced. Although agricultural losses typically dominate drought impacts, federal drought activities are not limited to agriculture. For example, the 2012 drought raised congressional interest in whether and to what extent other federal agencies have and are using authorities to address drought. Similarly, the President in August 2012 convened the White House Rural Council to assess executive branch agencies' responses to the ongoing drought. The Administration shortly thereafter announced several new administrative actions to address the drought. While numerous federal programs address different aspects of drought, no comprehensive national drought policy exists. A 2000 National Drought Policy Commission noted the patchwork nature of drought programs, and that despite a major federal role in responding to drought, no single federal agency leads or coordinates drought programs—instead, the federal role is more of "crisis management." Congress may opt to revisit the commission's recommendations. Congress also may consider proposals to manage drought impacts, such as authorizing new assistance to develop or augment water supplies for localities, industries, and agriculture—or providing funding for such activities where authorities already exist. Congress also may address how the two major federal water management agencies, the U.S. Army Corps of Engineers and the Bureau of Reclamation, plan for and respond to drought. This report describes the physical causes of drought, drought history in the United States, and policy challenges related to drought. It also provides examples of recurrent regional drought conditions.

In: Dams, Drought and Energy-Water Interdependencies ISBN: 978-1-62808-942-4
Editor: Daniel L. Calzi © 2013 Nova Science Publishers, Inc.

Chapter 1

DAMS AND ENERGY SECTORS INTERDEPENDENCY STUDY*

Tiffany Choi and Elizabeth Hocking

ABSTRACT

The U.S. Department of Energy (DOE) and the U.S. Department of Homeland Security (DHS) collaborated to examine the interdependencies between two critical infrastructure sectors – Dams and Energy.1 The study highlights the importance of hydroelectric power generation, with a particular emphasis on the variability of weather patterns and competing demands for water which determine the water available for hydropower production. In recent years, various regions of the Nation suffered drought, impacting stakeholders in both the Dams and Energy Sectors. Droughts have the potential to affect the operation of dams and reduce hydropower production, which can result in higher electricity costs to utilities and customers. Conversely, too much water can further complicate the operation of dams in ways that can be detrimental to hydropower production and to the infrastructure of the dams.

Discussions with dam owners and operators revealed that the storage capacity and conveyance flexibility of most conventional hydroelectric facilities were designed to accommodate local or regional historical patterns of hydrologic variability. Thus, episodic low water conditions, as opposed to long-term drought conditions, are not critical contributors to reduced hydropower production; however, the requirements for providing sufficient water for irrigation, environmental protection, transportation, as well as community and industrial uses are already in conflict in many places. Low water conditions (e.g., drought) and high water conditions (e.g., flood) resulting from extreme weather variability can strain the operation of dams and heighten the degree of competition for available water.

Although hydroelectric facilities are a type of asset that falls under the auspices of the Dams Sector, they are also an important element to the Energy Sector because the electric power they generate is critical to maintaining the reliability of the Nation's electricity supply. Therefore, this joint effort underscores the value of a cross-sector partnership model in the identification and discussion of issues significant to dam and

* This report was released by the U.S. Department of Energy and the Department of Homeland Securityon, dated September 2011.

utilities owners and operators, through which can help enhance their resilience against the potential impacts associated with the variability of weather patterns and extreme fluctuations of water flow.

SECTION 1: INTRODUCTION

The National Infrastructure Protection Plan (NIPP) provides an overarching framework for the protection and resilience efforts for the Nation's 18 critical infrastructure sectors.[2] Through the NIPP framework, each of the 18 sectors has developed public-private partnerships at an unprecedented level, providing a mechanism for critical infrastructure stakeholders to share cross-sector concerns and to collaborate on enhancing the protection and resilience posture of their critical infrastructure. This study complements the ongoing efforts of two critical infrastructure sectors—Energy and Dams—by examining the hydropower component of their close interdependency.[3]

The Department of Energy (DOE) and the Department of Homeland Security (DHS) are the designated Sector-Specific Agencies (SSAs) for the Energy and Dams Sectors, respectively. As the SSAs, DOE and DHS support and coordinate the protection and resilience activities for the Dams and Energy Sectors' critical infrastructure as defined below:

- Dams Sector assets include dam projects, hydropower generation facilities, navigation locks, levees, dikes, hurricane barriers, mine tailings and other industrial waste impoundments, and other similar water retention and water control facilities.[4]
- Energy Sector, as delineated by Homeland Security Presidential Directive 7 (HSPD-7), includes the production, refining, storage, and distribution of oil, gas, and electric power, except for hydroelectric and commercial nuclear power facilities.[5]

Although hydroelectric facilities are a type of asset that falls under the auspices of the Dams Sector, they are also an important element to the Energy Sector because the electric power they generate is critical to maintaining the reliability of the Nation's electricity supply. In preparing for this report, the SSAs for the Energy and Dams Sectors collaborated to examine the two sectors' shared concerns and interests in hydroelectric power generation. Chief among these concerns is the fact that hydroelectric power generation is affected by extreme fluctuations of water flow, as well as long-term issues surrounding the management and uses of water supply to generate hydroelectricity. In recent years, various regions of the Nation suffered droughts affecting stakeholders in both the Dams and Energy Sectors.[6] Although recent drought conditions have not caused a serious problem in terms of electricity supply and reliability, they have the potential to affect the operation of dams by decreasing hydropower production, which could result in higher electricity costs to utilities and customers.[7] Other weather-related variables such as air temperature, precipitation, and runoff conditions also impact future water supplies and demands, and may impose operational constraints on dams and utilities that rely on hydroelectric power generation.[8]

The report investigates how different variables might affect the operation of hydroelectric facilities and the supply of hydroelectric power, especially in times of drought and other extreme weather events. Such variables include:

- The relationship between hydroelectric power generation and the variability of hydrology and weather patterns;
- Operation of major reservoirs and streamflow regulations at these reservoirs; and
- Management for flood control, fish habitat protection, and power generation.

In addition, this joint effort underscores the value of the partnership model across sectors in the identification and discussion of the challenges and concerns that constitute priority issues for dam and utilities owners and operators. The ultimate goal of this effort is to help the two sectors enhance their resilience against the potential impacts associated with the variability of weather patterns and extreme fluctuations of water flow.

Limitations of the Study

To maintain the focus of the study, this report is limited to issues that specifically relate to electric power generation at hydroelectric dams. Specifically, this study examines issues pertinent to overall management of reservoirs and streamflows at dams that are affected by the variability of weather patterns. In-depth analysis of certain topics considered outside of the scope of the study is omitted from the report. These include: climate change, new hydropower technologies, renewable energy credits, the value of hydropower‘s avoided greenhouse gas emissions, and the effects of reduced hydropower generation on the overall power market.

There are three types of hydroelectric power plants: conventional, pumped storage, and diversion facilities. The focus of this report is on the conventional hydroelectric facilities, which are the most common type of hydroelectric power plant.[9] The U.S. Energy Information Administration (EIA) defines a conventional hydroelectric power plant as a plant in which all of the power is produced from natural streamflow as regulated by available storage.[10] Most pumped storage units have closed-loop systems in which water can be stored and reused; therefore, electricity production at pumped storage is more resistant to drought or changing weather patterns. For this reason, the discussion of and data on hydroelectric power generation provided in this report excludes generation from pumped storage, unless noted otherwise.

While the operation of thermoelectric plants is significantly affected by the availability of water, they are not the subject of this report. A 2009 report from Sandia National Laboratories – New Mexico entitled, *Energy and Water Sector Policy Strategies for Drought Mitigation,* examined the use of water in the electricity production process and how different technologies can affect a plant‘s water requirements and raise environmental concerns.[11] Appendix E provides a brief summary of this discussion from the Sandia report.

Finally, it should be noted that there is a tremendous amount of ongoing and proposed activity relating to hydroelectric power, as well as broader water management and supply issues.

It is not the purpose of this study to consider or catalogue all such efforts. Appendix F provides references related to dams and hydroelectricity that may provide further background information.

SECTION 2: HYDROELECTRIC POWER IN THE UNITED STATES

This section provides an overview of hydroelectric power generation in the United States to demonstrate the significance of hydroelectric dams in the Energy Sector. The following subsections provide a national overview of hydroelectric power generation, including its key benefits, historical capacity and generation data, the variability of weather and hydropower production, recent changes to hydropower generation, and a list of the 20 largest hydroelectric dams in the United States.

2010 U.S. Hydropower Facts

- Hydroelectric sources produce seven percent of the U.S. total annual electric generation.
- Hydroelectric generating capacity constitutes eight percent of the U.S. total existing generation capacity.
- Top ten hydropower-generating States produce more than 80 percent of the U.S. total hydroelectric generation.
- The 20 largest hydroelectric dams produce almost half of the U.S. total hydroelectric generation.
- Hydroelectric power generation has declined in most parts of the country during the 2007-2009 period compared to the historical average.

2.1. Importance of Hydroelectric Dams for Power Generation

Historically, hydroelectric sources have been a vital source of electric power generation that accounted for as much as 40 percent of the Nation‘s electricity supply in the early 1900s.[12] Although the share of hydropower generation has declined to seven percent of the U.S. total electric power generation as production from other types of power plants grew at a faster rate, hydroelectric dams remain an important power source.[13]

Hydropower is critical to the national economy and the overall energy reliability because it is:

- The least expensive source of electricity, as it does not require fossil fuels for generation;
- An emission-free renewable source, accounting for over 65 percent of the U.S. total annual net renewable generation;[14]
- Able to shift loads to provide peaking power (it does not require ramp-up time like combustion technologies); and
- Often designated as a black start source that can be used to restore network interconnections in the event of a blackout.

Hydroelectric power is derived from the force of moving water. It is considered a "renewable" source, because the water on the earth is continuously replenished by precipitation.[15] A typical hydro plant serves multiple functions and consists of three parts: a power plant where the electricity is produced, a dam that can be opened or closed to control water flow, and a reservoir where water can be stored. [16] The water behind a dam flows through an intake and pushes against blades in a turbine, causing them to turn and produce electricity. The amount of electricity that can be generated depends on how far the water drops and how much water moves through the system.

In addition to providing clean electricity production, hydropower serves an essential purpose of enhancing electric grid reliability. Hydropower can rapidly adjust output to meet changing real-time electricity demands and provide "black start" capability to help restore power during a blackout event. Black start capability is defined as the ability to start generation without an outside source of power.[17] Because hydropower plants are the only major generators that can dispatch power to the grid immediately when all other energy sources are inaccessible, they provide essential back-up power during major electricity disruptions such as the 2003 blackout.[18] With black start capability, hydropower facilities can resume operations in isolation without drawing on an outside power source and help restore power to the grid.

Figure 1 is a snapshot of hydropower generation in the United States today. The 10 highlighted States together produce more than 80 percent of the Nation‘s total hydroelectric power. The numeric values represent each State‘s dependence on hydro sources for electricity generation. For example, hydro sources in Maine, South Dakota, and Vermont each contribute less than two percent of the Nation‘s hydroelectric generation; however, their dependence on hydro sources is relatively high. Conversely, Alabama, Arizona, and North Carolina each produce more than two percent of the Nation‘s hydroelectricity, but their reliance on hydro sources for electric power generation is relatively low.

2.2. Hydroelectric Power Capacity vs. Generation

As seen in figures 2 and 3, hydropower generation capacity has remained steady in the last 20 years, whereas production from hydro sources has fluctuated dramatically year-to-year. According to EIA, hydropower capacity grew at an annual rate of 0.3 percent or a total of 4,600 megawatts (MW) in the past 20 years (1990: 73,925 MW vs. 2009: 78,525 MW).[19] EIA projects a minimum growth in hydroelectric generation capacity (0.1 percent annual rate) and a slightly greater increase in hydropower generation, with an annual growth rate of 0.5 percent over the next 25 years.20 Despite these forecasts, it is almost impossible to predict the interannual variability of hydropower generation in the United States because the operation of hydroelectric facilities is directly linked to the amount of precipitation received.

2.3. Variability of Weather and Hydroelectric Power Generation

Hydroelectric power generation depends on the availability of local water sources that are susceptible to changes in local hydrology and weather patterns. Operational policies (i.e., flood control as the primary mission) and regulatory compliance (e.g., instream flow requirements for fish protection) are important factors in hydropower generation, as are multiple competing water uses such as water supply, irrigation, and recreation.

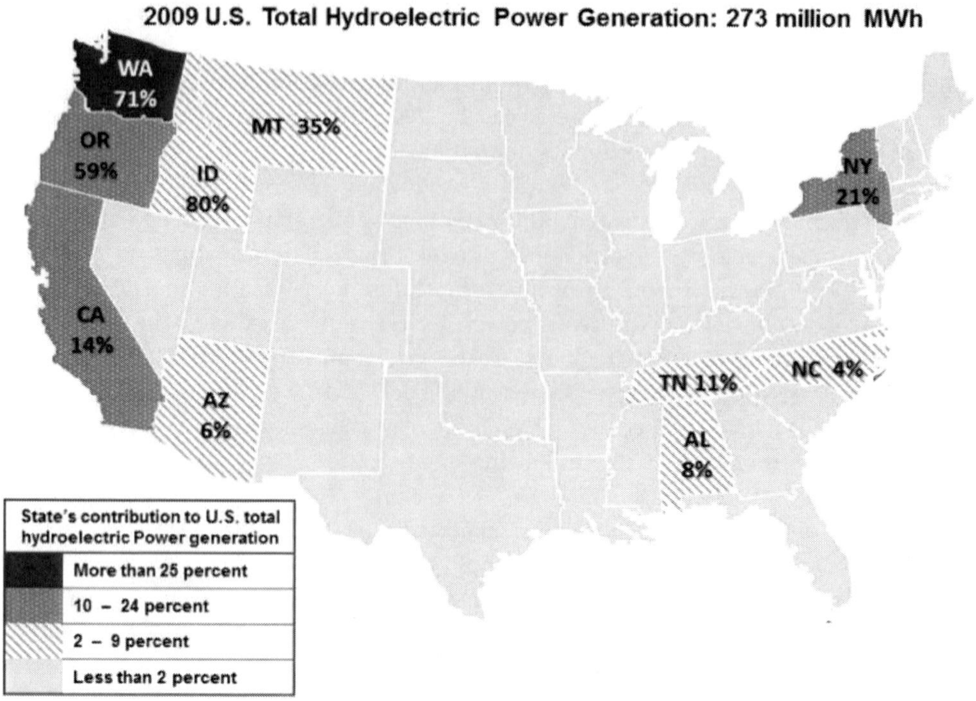

Source: Derived from EIA-906, EIA-920, EIA-923 databases, http://www.eia.doe.gov/cneaf/electricity/page/eia906 920.html, accessed November 30, 2010.

Note:

- Highlighted States represent top 10 States that generate the most hydroelectricity as color-coded.
- The numeric value in each State represents the share of hydro sources in that State's total power generation.
- For example, Washington State's hydro sources generate over 25 percent of the Nation's total hydroelectric power generation and 71 percent of State's total electric power generation.

Figure 1. Top 10 Hydropower-Generating States and Their Reliance on Hydro Sources for Electricity, 2009.

In other words, the operation of a hydroelectric facility is affected by the amount of water available in a river basin where the facility is located, as well as competing uses of water that are specific to each river.[21] (See section 3 for discussion on the operational and regulatory issues impacting water uses and streamflows.)

In the past century, total precipitation has increased by about seven percent averaged across the United States.[22] However, year-to-year fluctuation in natural weather and climate patterns can produce a period that does not follow the long-term trends (see figure 3). The interannual variability of hydropower generation in the United States is very high—a drop of 59 million megawatt hours (MWh) (or 21 percent of the U.S. total hydropower generation) was seen from 2000 to 2001. Sensitivity of hydroelectric power generation to changes in precipitation and river discharge is high; in the range of 1.0+ (a sensitivity level of 1.0 means that one percent change in precipitation results in one percent change in generation).[23]

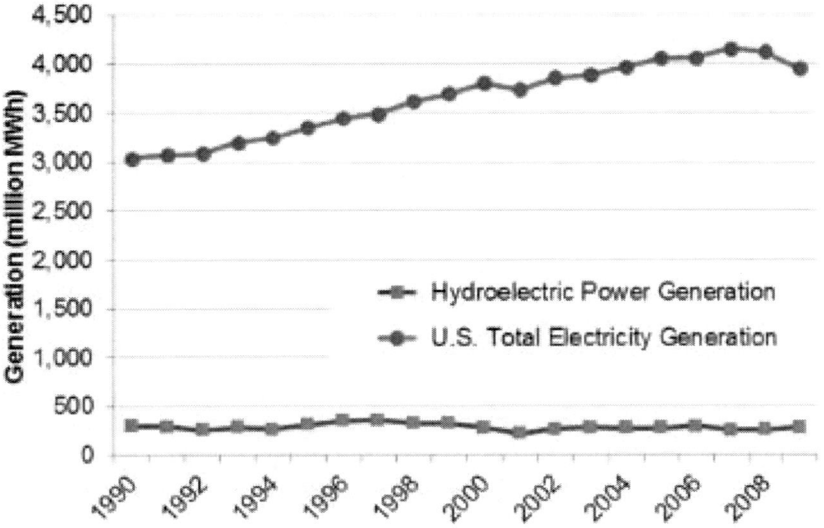

Figure 2. Electricity Generation Growth, 1990-2009.

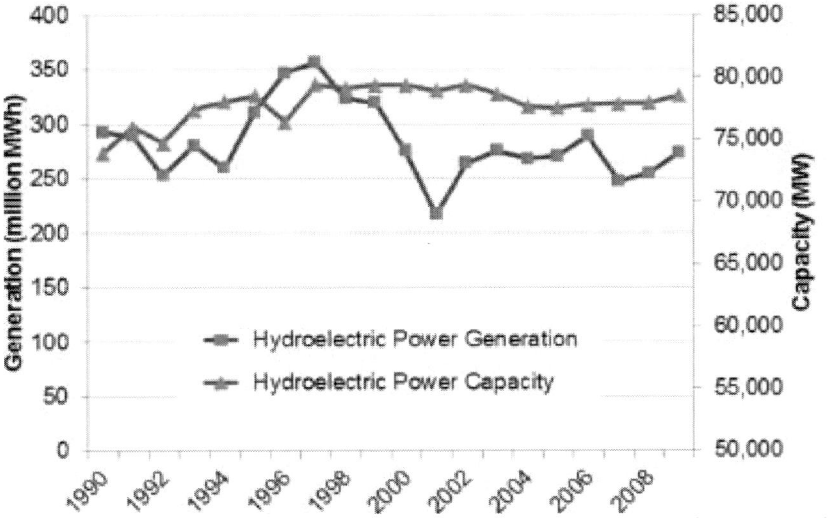

Sources for both figures: Derived from EIA-906, EIA-920, EIA-923, and EIA-860 databases, http://www.eia.doe.gov/cneaf/electricity920.html and http://www.eia.doe.gov/ cneaf/electricity /page/eia860.html, accessed November 28, 2010.

Figure 3. Comparison of Hydropower Capacity vs. Generation, 1990-2009.

Although it is evident that precipitation is a determining factor in available hydropower generation for a given period of time, the variability of weather patterns impose uncertainty in the operation of hydroelectric facilities.

Hydropower operations are also affected indirectly by the changes in air temperatures, humidity, and wind patterns which change water quality and reservoir dynamics.[24] For example, reservoirs with large surface areas (such as Lake Mead in the lower Colorado River) are more likely to experience greater evaporation, which affects the availability of water for all uses including hydropower. In addition, altering snowfall patterns and associated runoff from snowpack melt are a matter of concern, particularly in the Pacific Northwest, where snows are melting earlier and the proportion of precipitation in the form of snow is decreasing.[25]

2.4. Historical Hydroelectric Power Generation

The dependability of hydroelectric power generation is often challenged by unusual and frequently unpredictable weather patterns including droughts, floods, and early snowpack melts. Lower streamflows resulting from drought, upstream dams, and diversions will reduce the amount of storage in a reservoir which lowers the amount of water that can be used to produce hydropower. Coupled with operational constraints under certain streamflow requirements, the diminished streamflows can reduce hydropower production. Such decline may complicate electricity providers‘ ability to meet their power supply commitments, especially in service areas that depend heavily on hydroelectric power. However, reduced hydropower generation caused by regional drought conditions may often be replaced by increased fossil fuel-based generation.

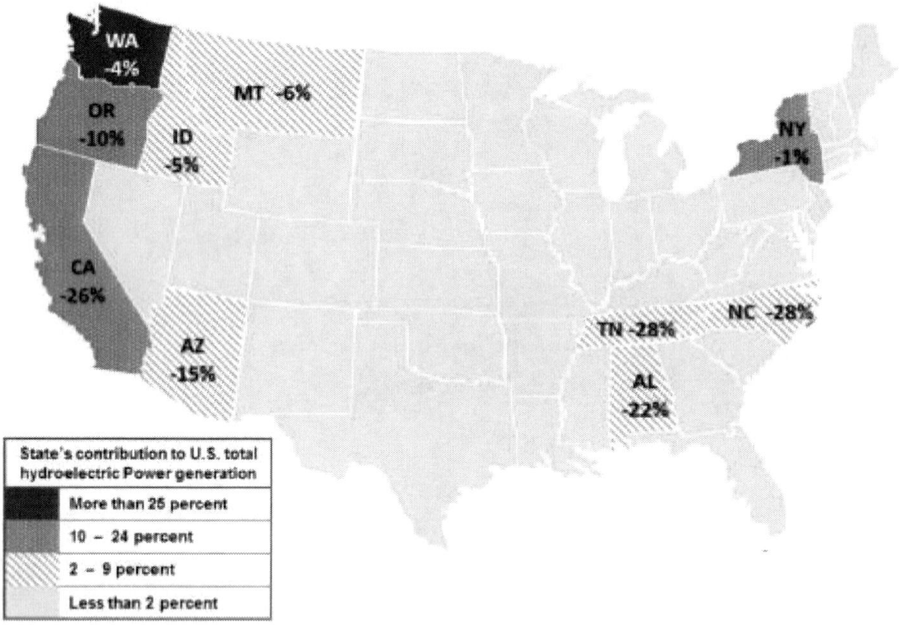

Source: Derived from EIA-906, EIA-920, EIA-923 databases, http://www.eia.doe.gov/cneaf/electricity/page/eia906 920.html, accessed November 28, 2010.

Note: Figures are calculated as percent change in hydropower generation between 2007 and 2009 in comparison to the historical average from 1990 to 2006.

Figure 4. Variance in Annual Average Hydropower Generation in Top 10 States, 2007 2009 vs. Historical Average.

Figure 4 shows recent changes in hydroelectric power production in the top ten hydropower-generating States. A 20-year period from 1990 to 2009 was examined to see the changes in hydropower production at the State level. The results indicate that the national annual average of hydroelectric power generation between 2007 and 2009 was 11 percent less than that of the historical average between 1990 and 2006. As seen in this figure, all top 10 hydropower-producing States experienced a decline, with certain States losing up to 28 percent of their normal annual hydropower generation.

2.5. Largest Hydro Dams

According to the 2010 Dams Sector-Specific Plan, the total number of dams in the United States is estimated to be around 100,000. However, most dams were constructed solely to provide irrigation and flood control, and only about two percent (or 2,000) of the Nation's dams produce electricity.[26] Approximately half of U.S. hydropower generation capacity is federally owned and operated (e.g., owned by the U.S. Army Corps of Engineers (USACE), Bureau of Reclamation (Reclamation) of the U.S. Department of Interior (DOI), and Tennessee Valley Authority (TVA)); the other half consists of nonfederal projects that are regulated by the U.S. Federal Energy Regulatory Commission.

Table 1 provides a list of the 20 largest hydroelectric dams in the United States ranked by summer capacity as of December 2009. These 20 hydroelectric facilities account for 40 percent of the Nation's hydroelectric power capacity; they provided 44 percent of the hydropower generated in the United States during the 20-year period from 1990 to 2009. The majority of the 20 largest hydroelectric power plants are located in the Columbia River basin in the Pacific Northwest, all of which experienced decreased production in the 2007 to 2009 time span compared to the historical average between 1990 and 2006.[27] EIA reports that the largest hydroelectric facility in the United States is the Grand Coulee Dam with a summer capacity of 6,765 MW, located in the Columbia River basin.[28] It is also the largest hydropower producer, generating about eight percent of the Nation's hydropower. To compare the magnitude of the Grand Coulee, the next two largest dams, Chief Joseph and Robert Moses Niagara, each have only about a third of Grand Coulee's capacity. Note, however, that the capacity factor at hydro plants varies significantly, generally in the range of 30 to 80 percent, with an average capacity factor of about 40 to 45 percent.[29] To illustrate this varied capacity factor of hydroelectric plants, the capacity factor of the Grand Coulee Dam is about 36 percent, whereas the Robert Moses Niagara Dam has a relatively high capacity factor of 71 percent.[30]

SECTION 3: OPERATION OF HYDROELECTRIC DAMS IN SELECTED MAJOR WATERSHEDS

The operation of a hydroelectric power plant is subject to various internal and external factors. Internal factors include the management of hydro dams, which respond to the upstream and downstream conditions by controlling the volume and timing of water retained or released. External factors include constraints imposed by alternative uses of water (navigation, irrigation, water supply, fish habitat, recreation) that may lead to restricted flow rates.[31]

Table 1. 20 Largest Hydroelectric Dams in the United States

Plant Name	Owner	State	Initial Operating Year		Summer Capacity (MW)	Comparison of HistoricalAvg. Annual Generation (MWh)		Difference in Avg. Gen. (%) 2007-2009 vs. 1990-2006	Capacity Factor
					2009	1990-2006	2007-2009		2009
Grand Coulee	USBR	WA	1941	6,765	21,170,076	21,596,413		2%	35%
Chief Joseph	USACE	WA	1955	2,456	11,454,051	10,684,406		-7%	45%
Robert Moses Niagara	NYPA	NY	1961	2,353	14,543,029	14,021,163		-4%	71%
John Day	USACE	OR	1969	2,160	9,958,204	8,703,430		-13%	44%
Hoover Dam	USBR	AZ-NV	1936	2,079	4,429,576	3,723,415		-16%	20%
The Dalles	USACE	OR	1957	1,823	6,988,641	6,320,308		-10%	38%
Glen Canyon Dam	USBR	AZ	1964	1,312	4,297,797	3,680,952		-14%	32%
Rocky Reach	PUD	WA	1961	1,254	6,003,149	5,808,323		-3%	49%
Bonneville	USACE	OR	1938	1,093	4,919,740	4,503,497		-8%	47%
Wanapum	PUD	WA	1963	1,044	5,019,250	4,790,141		-5%	39%
Boundary	Seattle	WA	1967	1,040	3,861,324	3,674,757		-5%	48%
McNary	USACE	OR	1953	991	6,061,311	5,211,778		-14%	59%
Priest Rapids	PUD	WA	1959	932	4,605,956	4,642,458		1%	52%
Wells	PUD	WA	1967	840	4,303,039	3,963,250		-8%	51%
Lower Granite	USACE	WA	1975	810	2,479,234	2,042,004		-18%	34%
Little Goose	USACE	WA	1970	810	2,423,408	2,056,557		-15%	33%
Lower Monumental	USACE	WA	1969	810	2,502,151	2,099,973		-16%	33%
Robert Moses Power Dam	NYPA	NY	1958	800	6,771,500	6,936,087		2%	90%*
Oahe	USACE	SD	1962	714	2,353,409	1,355,022		-42%	30%
Shasta	USBR	CA	1944	714	1,800,872	1,605,009		-11%	23%
Total20 Dams				31,113	125,945,718	117,418,942		*-7%*	
Total U.S.				78,518	287,855,374	256,708,605		*-11%*	
20 Largest Dams as percent of U.S. TotalHydro				*40%*	*44%*	*46%*			

Note:

- This table compares the historical annual average generation (1990-2006) with that of the recent three years, 2007-2009, at the 20 largest dams in the United States, ranked by summer capacity.
- Initial operating year represents the year in which the first unit(s) at the plant became operational and does not document the years in which additional units were brought online at the same facility.
- Owner information:

- NYPA: New York Power Authority
- PUD: Public Utility District
- Seattle: Seattle City of Light
- USACE: U.S. Army Corps of Engineers
- USBR: U.S. Bureau of Reclamation

- Capacity Factor: calculated using the 2009 summer capacity and generation data, except for Robert Moses Power Dam, for which the nameplate capacity of 912MW is used. Source: Derived from EIA-906, EIA-920, EIA-923, and EIA-860 databases, http://www.eia.doe.gov/cneaf/electricity/page/eia906 920.html and http://www.eia.doe.gov/cneaf/electricity/page/eia860.html, accessed November 28, 2010.

Available water flow is significant both as an internal factor, if upstream and basin flows decrease, and as an external factor, if water management requires distribution of water for other purposes or maintaining water to support reservoir activities. This explains why drought can play a significant role in hydropower production—it can decrease upstream flow and require the diversion or retention of water that would otherwise go to produce electricity or to other water purposes during times of scarcity.

The operations of a river system and of hydroelectric plants on that river are guided by a set of complex rules, policies, and agreements that vary vastly by the location and functions of each river system. While certain Federal laws may apply to all major watersheds, there are numerous State and local laws that specifically govern each river. In addition, river systems that cross national borders are subject to international policies and agreements. In other words, each watershed faces distinct issues and policies. To explore these unique factors, this section provides a brief overview of three major river systems—the Columbia River, the Colorado River, and the Tennessee River—and the various issues affecting the hydro dam functions and operations in each watershed.

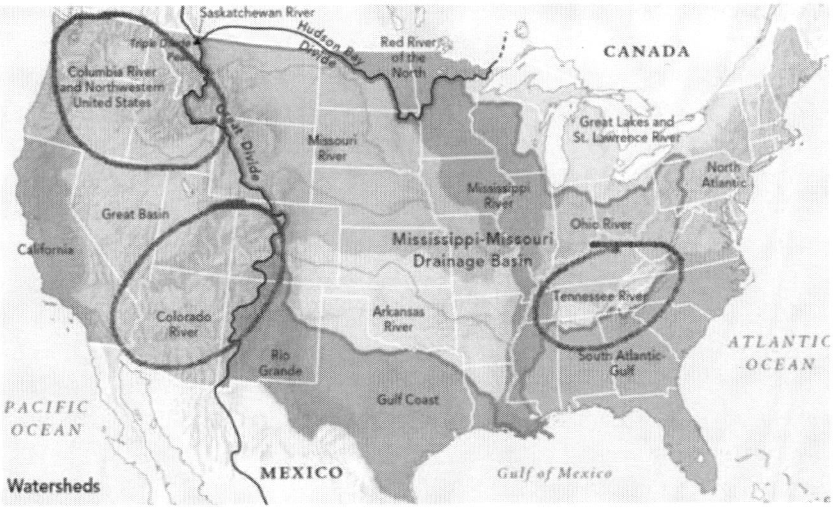

Source: http://maps.howstuffworks.com/united-states accessed January 4, 2010.

Figure 5. Selected Major Watersheds in the U.S.

3.1. The Columbia River System

The Columbia River basin is the predominant river system in the Pacific Northwest, encompassing 250 reservoirs and about 150 hydroelectric projects.[32] The system spans seven western States: Washington, Oregon, Idaho, Montana, Wyoming, Nevada, and Utah, as well as British Columbia, Canada (see figure 6).[33] USACE and Reclamation are the owners and operators of the 31 federally-owned hydro projects in the river system; the Bonneville Power Administration (BPA) markets and distributes power generated at Federal dams in the region.[34]

Today, the Columbia River system operations serve multiple purposes—flood control and mitigation, power production, navigation, recreation, and environmental needs—that are guided by a complex and interrelated set of laws, treaties, agreements, and guidelines. These include the Endangered Species Act, a Federal law that protects threatened or endangered species protection that can result in setting restrictions on the time and amount of allowed flow and spill—as well as numerous treaties and agreements with Canada dealing with flood control and division of power benefits and obligations.[35]

Streamflow in the Columbia River system does not follow the region‘s electricity demand pattern in which the peak occurs during winter when the region‘s homes and businesses need heating. Although most of the annual precipitation occurs in the winter from snowfall, most of the natural streamflows occur in the spring and early summer when the snowpack melts. About 60 percent of the natural runoff occurs during May, June, and July (see figure 7). Thus, the objective of reservoir operation is to store snowmelt runoff in the spring and early summer for release in the fall and winter when streamflows are lower and electricity demand is higher.

Hydropower Operation and Planning

Hydropower supplies approximately 60 to 70 percent of the electricity in the Pacific Northwest Region.[36] In the Columbia River system, power generation operations are generally compatible with flood control requirements. However, under the current operating strategy, conflicts between power generation and fish protection are generally resolved in favor of fish protection.

The current strategy requires increased water storage in the fall and winter and increased flows and spill during the spring and summer to benefit migrating juvenile salmon. This approach does not provide an optimal operating strategy for power generation as it results in more water for fish protection, but reduced hydropower generation during the peak demand periods.

As a result, BPA is often likely to purchase power frequently during high load periods in the winter and sell surplus power in the spring and summer. (See figure 8 for critical rule curves applied at a typical Columbia River reservoir). According to Steve Wright, administrator of BPA, his agency has reduced output of Federal hydropower by about 1,000 MW as a result of protections to restore threatened and endangered salmon and steelhead over the past 20 years.[37]

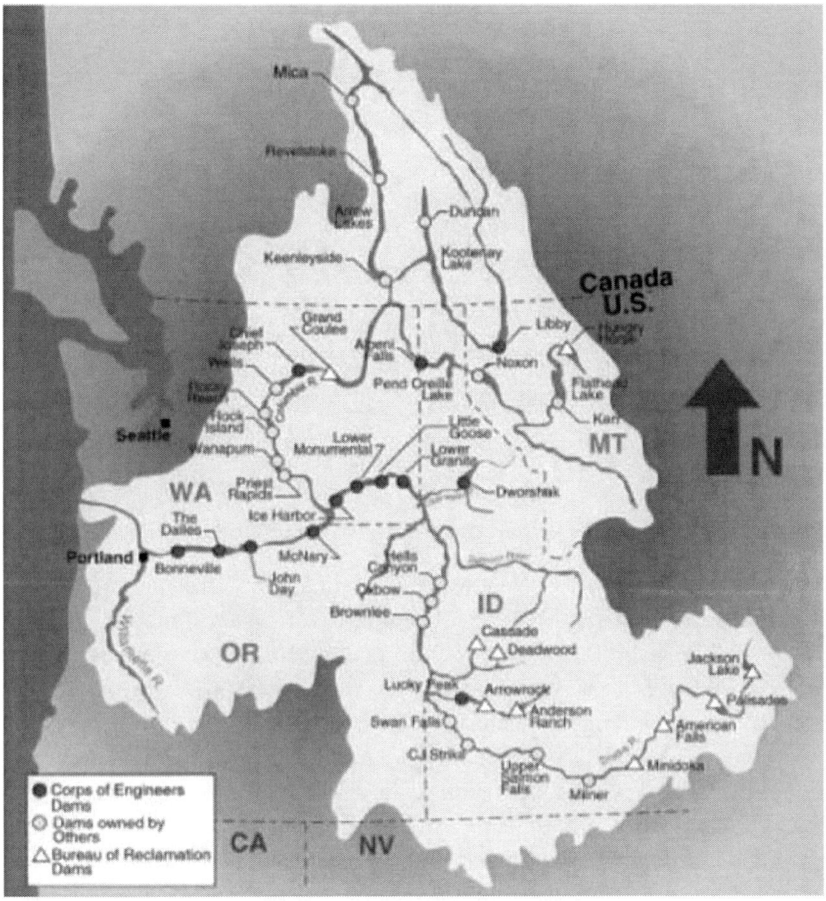

Source: Columbia River Basin Map, NOAA, http://www.nwr.noaa.gov/Salmon-Hydropower/Columbia Snake-Basin/upload/Col-Basin-map.pdf, accessed December 12, 2010.

Figure 6. Columbia River Basin.

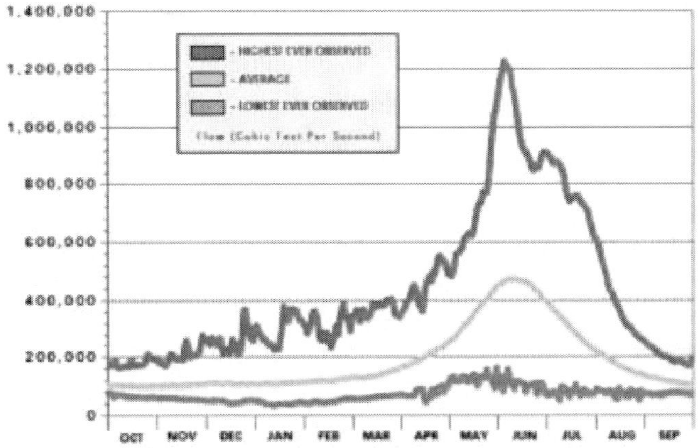

Note: Flow on the Columbia River is measured at the Dalles, Oregon.

Figure 7. Columbia River Streamflows.

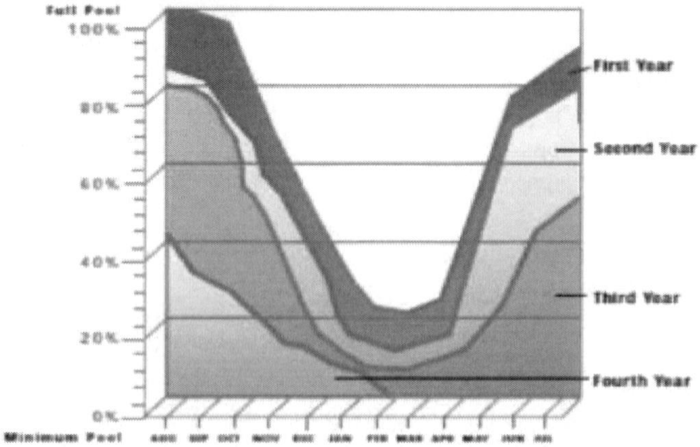

Source for both figures: BPA, http://www.bpa.gov/power/pg/columbia river inside story.pdf, accessed December 16, 2010.

Figure 8. Critical Rule Curves for a Typical Columbia River Reservoir.

Two agreements, the Pacific Northwest Coordination Agreement (PNCA) and the Columbia River Treaty, underpin how the Columbia River system functions in a coordinated fashion. The Columbia River Treaty enables improved water storage and annual planning for river projects with Canada, from which 25 percent of the streamflow originates. The PNCA directs the coordination among the Federal project operators and hydroelectric generating utilities in the region. The PNCA enables the optimization of system reliability and power production, provided that it is consistent with requirements for nonpower uses or functions.[38]

The PNCA Coordinating Group, made up of BPA, USACE, Reclamation, and major generating utilities in the Pacific Northwest and Canada, oversees planning and operation for power production. Annually, the group develops a set of operating guidelines called —operating rule curves" to guide reservoir operations for power production. Such planning is based on the possibility that the lowest historical streamflow conditions (—four-year critical period" from 1928 to 1932) could recur (see figure 8). The guidelines also include a flood control curve that requires an adequate space in the reservoir to regulate the predicted runoff for the year without causing flooding downstream.

3.1.1. Effects of Changing Streamflow on Hydroelectric Power Generation

The Pacific Northwest has been affected by widespread temperature-related reductions in snowpack, as well as a changing annual runoff pattern. Recent studies indicate 1) a transition to more rain and less snow[39] and 2) a shifting pattern of snowmelt runoff in western North America— contemporary snowmelt runoff has been observed 10 to 30 days early in comparison to the period from 1951 to1980 (see figure 9).[40] To adapt to these changes, the ability to modify operational rules and water allocations is critical to ensuring the reliability of water and energy supplies, as well as to protecting the environment and critical infrastructure. However, the current set of laws, regulations, and agreements is intricate and creates institutional and legal barriers to such changes in both the short and long term.[41]

In 2010, the Pacific Northwest experienced the third driest year in the last 50 years and the fifth lowest water level on record since 1929, causing low runoff in the lower Columbia River.[42] According to BPA's 2010 Annual Report, BPA's gross purchased power increased

$104 million, or 37 percent, from 2009, mainly due to below normal basin-wide precipitation and streamflows, resulting in insufficient power generation to fulfill load obligations.[43]

As a result, BPA experienced a net loss of $233 million, or 10 percent, from the prior year due to reduced hydropower generation.

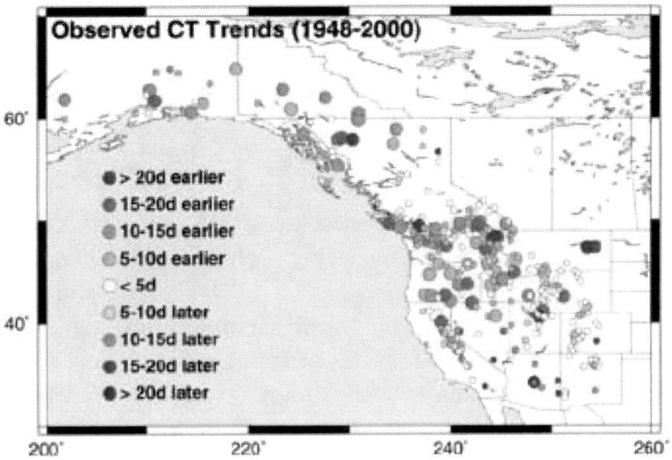

Source: Stewart et al. (see footnote 40).

Figure 9. Observed Changes in Timing of Center of Mass of Flow (CT), 1948-2000 (Reference Time Period: 1951-1980).

Despite the below-normal hydroelectric generation, electricity supply to customers remained adequate in the Pacific Northwest and rates were unaffected. The rates to customers were unaffected because the rates BPA charges customers were locked in for two years. However, the rates are expected to go up by five to six percent in 2011 when contracts are reviewed due to uncertainties around continuing low water supplies as well as ongoing litigation over salmon conservation in the lower Columbia River basin that is intricately tied to hydropower operation.[44]

Not only droughts, but too much water can also bring challenges to hydropower operation. After a dry winter, spring 2010 river flows were expected to stay fairly low. However, in June 2010, a strong Pacific storm system brought heavy precipitation that almost doubled the streamflows in the Columbia River.[45] During the month of June, dam operators faced the challenges of managing flooding and an oversupply of hydropower and, at the same time, complying with Federal regulations for fish protection that restricted the amount of spill allowed. Since water that conditions for fish, dam operators were forced to produce power for which they could not find a market.[46] As a result, BPA disposed of more than 50,000 MWh of electricity for free or for less than the cost of transmission and incurred a total of 745,000 MWh of spill for lack of market in June 2010.[47] Figure 10 shows that BPA balancing authority generation significantly exceeded load in early June.

High flows in the Columbia River system are common, resulting from above average snowpack and/or early warming periods that result in rapid snowmelt. However, operating the Columbia River system through those events has become much more complex in recent years due to the following new factors: 1) multiple flow and storage requirements to protect threatened and endangered salmon and steelhead under the Endangered Species Act; 2) changing uses of the transmission system in a deregulated electric power market; and 3) the significant addition of variable, non-dispatchable wind power capacity (3,400 MW as of February 2011) with financial incentives for operation—production tax credits of $21 per MWh and renewable energy credits of $20 per MWh.[48]

The oversupply of hydropower in a statistically low water year demonstrates the limitations of cumulative statistics and the challenges of managing the Columbia River system. Such conditions can be exacerbated in a heavy water year, especially with the forecasted interconnection of an additional 3,000 MW wind generation capacity to BPA's system over the next few years.[49]

In response to these challenges, BPA has ongoing efforts to address excess supply of energy that can cause physical and operational constraints on Federal hydro and transmission system operations.[50] In May 2011, BPA adopted the Final Record of Decision on the Interim Environmental Redispatch and Negative Pricing Policy (Interim Policy) that would be implemented during high water and high wind events. [51] The purpose of the Interim Policy is to assure that the system is able to comply with environmental mandates while allowing reliable and equitable power production in the BPA balancing authority area. It proposes to achieve this by 1) limiting generation at coal, natural gas, and other thermal power plants to keep the supply of power from exceeding demand and 2) allowing temporary curtailment of wind generation connected to its power transmission system. Furthermore, BPA would not pay negative prices if it needed to generate electricity to meet environmental requirements. BPA continues to work with its regional partners to seek long-term solutions as the Interim Policy is set to expire after March 30, 2012.

3.2. The Colorado River System

The Colorado River is considered one of the most legally complex river systems in the world, governed by multiple interstate and international compacts, legal decrees, and prior appropriation allocations, as well as federally-reserved water rights for Native Americans.[52] The river basin extends over seven U.S. States—Arizona, California, Colorado, Nevada, New Mexico, Utah, and Wyoming and parts of northwestern Mexico (see figure 11), serving about 25 million people in the Southwest. Its water yield is only eight percent of the annual flow of the Columbia River, yet it is arguably the most regulated river in the country.[53]

The river is governed by the —Law of the River" that consists of the 1922 Colorado River Compact (Compact) and the 1948 Upper Colorado River Basin Compact, along with the 1944 International Treaty with Mexico, a number of Federal laws, and U.S. Supreme Court decisions.[54]

The river is divided into two areas, upper Colorado and lower Colorado, and water is directed to be allocated equally between the two regions based on historical rainfall patterns. However, the Compact that regulates the water allocation is believed to have been negotiated in a period of abnormally high rainfall, resulting in allocation of water greater than the sustainable quantity.[55]

Consequently, the river has been the source of disputes among States, between the United States and Mexico, between cities and farms, between power users and conservationists, and between Indian tribes and non-Indian water users.[56]

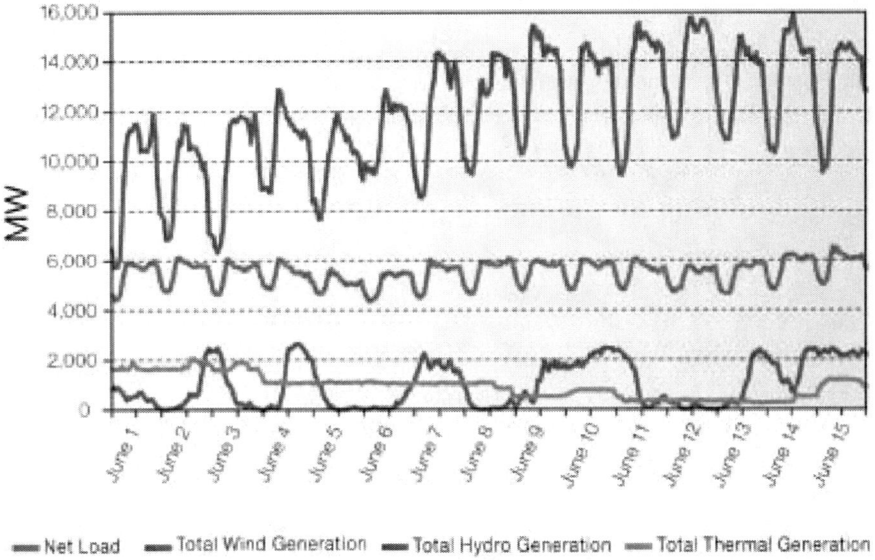

Source: June 2010 High Water Report, BPA (see footnote 45).

Figure 10. BPA Balancing Authority Load, June 1-15, 2010.

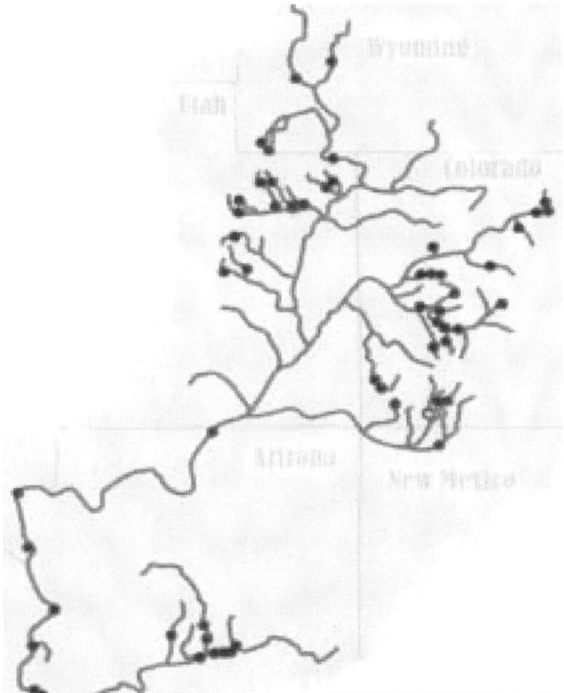

Source: http://www.pbs.org/cowboysindianslawyers/top icfeature.html, accessed December 12, 2010.

Figure 11. Colorado River Basin.

3.2.1. Effects of Droughts on Hydroelectric Power Generation

In the early 21st century, water use issues intensified as the Colorado River region experienced some of the Nation‘s highest population growth, as well as the start of a long period of drought considered to be the worst drought in the 100-year recorded history (hereinafter referred to as the —early 21st century drought").[57] (See figures 12, 13, and 14.) The Colorado River region is of particular concern because of the continuing trend of rising temperatures seen across the region that contributes to increased evaporative losses from snowpack, surface reservoirs, irrigated land, and vegetated surfaces.[58]

Although certain temperature trends are evident, the projections of future precipitation remain unclear, leading to uncertainty in possible changes in future streamflow in the Colorado River.

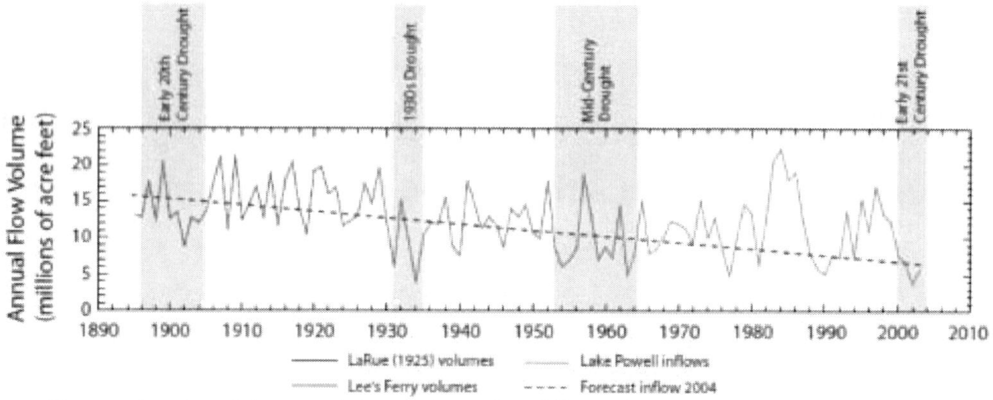

Source: Time-series plot of the annual flow volume (in millions of acre-feet) for the Colorado River at Lee‘s Ferry, USCG, http://wwa.colorado.edu/colorado river/docs/USGS 2004 3062.pdf, accessed December 12, 2010.

Figure 12. Annual Flow Volume for the Colorado River at Lee's Ferry.

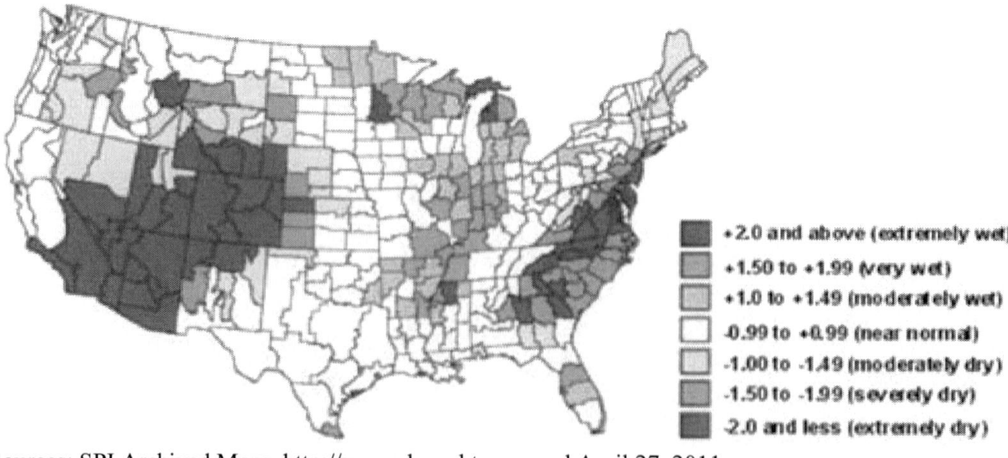

Sources: SPI Archived Maps, http://www.drought accessed April 27, 2011.

Figure 13. 12-Month Standardized Precipitation Index (SPI), Sept. 2001-Aug. 2002.

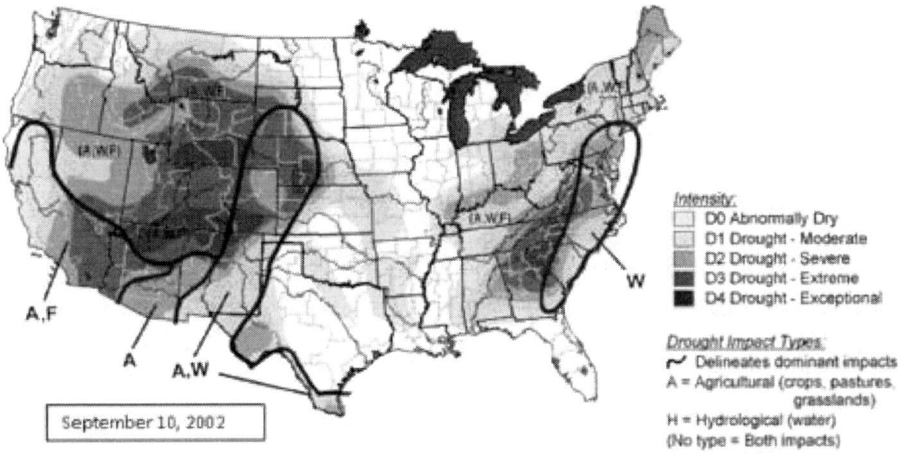

Source: U.S. Drought Monitor Archives, http://www.drought.unl.edu/dm/archive.html, accessed January 9, 2011.

Figure 14. Drought Monitor Map, Week of September 10, 2002.

A number of activities have been underway to cope with drought in the Colorado River region. The Colorado River basin States have engaged in long-range water planning, drought management, and conservation measures.[59] In 2005, Reclamation launched an effort to develop strategies for improving coordinated management of the two largest reservoirs in the river, Lakes Mead and Powell, during drought and low reservoir conditions. Lakes Mead and Powell comprise approximately 80 percent of the basin‘s entire storage capacity.[60]

The Reclamation-led effort resulted in the development of the *Colorado River Interim Guidelines for Lower Basin Shortages and Coordinated Operations for Lake Powell and Lake Mead* (Interim Guidelines) that sought to address the unique challenges in the operation of the Colorado River. [61] This effort included an environmental review and public scoping meetings during which many stakeholders raised concerns relating to dam operations during drought and low reservoir conditions. However, some stakeholders expressed a need to consider other water supply, water management, and operational strategies or programs that could improve the availability and reliability of Colorado River water supplies.

Some of the most frequently raised comments during an environmental impact statement (EIS) process included:

- Consider/evaluate costs and benefits of decommissioning the Glen Canyon Dam,
- Consider/evaluate transfer of Lake Powell and Lake Mead storage to groundwater aquifers, and
- Update the Compact to reflect the Colorado River‘s supply limitations and changing societal demands.[62]

In December 2007, a record of decision was issued, officially adopting the Interim Guidelines, including four new operational rules and guidelines that:

- Establish conditions for shortages, specifying when and who will take reductions in the allocated water (this is essential for prudent water planning in times of drought);

- Allow the water level in Lake Powell and Lake Mead to rise and fall in tandem, thereby better sharing the risk of drought;
- Allow DOI to allocate surplus water should there be abundant runoff in the basin; and
- Address the ongoing drought by encouraging new initiatives for water conservation.[63]

Figure 15 shows the historical elevation level at Lakes Mead and Powell. In October 2010, Lake Mead stood at 39 percent capacity or 1,084 feet in elevation, curtailing power generation at the Hoover Dam, the region's largest hydro facility. For every foot of elevation lost in Lake Mead, Hoover Dam produces 5.7 MW less power. That is because at lower water levels air bubbles flow through with the water causing the turbines to lose efficiency.[64] As a result, electricity available from Hoover Dam declined 29 percent since 1980, which meant that local utilities had to buy power on the open market where rates were up to four times higher.[65] (Also see table 1 in section 1 for the hydroelectric generation at Hoover Dam between 1990 and 2009.)

Although it is uncertain when this drought will come to an end, the condition of the Colorado River improved significantly in summer 2011. In June 2011, Reclamation reported that the water level at Lake Powell increased significantly as a result of a long and wet winter. The reservoir elevation of Lake Powell was projected to reach 35 to 40 feet below the full elevation of 3,700 feet by August 2011; a level last seen in October of 2001, near the beginning of the early 21st century drought. [66] However, the storage in Lake Powell was still considered below the desired operating level.[67]

The Federal Government has ongoing efforts to address issues related to drought. Currently, Reclamation is conducting the *Colorado River Basin Water Supply & Demand Study* that will characterize current and future water supply and demand imbalances in the basin.[68] The study will also develop and analyze adaptation and mitigation strategies to resolve those imbalances, and is scheduled to conclude in July 2012.

In addition, Reclamation issued a $3.4 million contract to upgrade generating facilities at the Hoover Dam in April 2010.69 Per the contract, a new turbine will be installed to allow the generating units to operate more efficiently over a wide range of water levels, enabling Hoover Dam to generate power at lake levels as low as 1,000 feet. The new turbine is expected to be delivered in early 2012 and, if successful, several of the other 16 turbines would be replaced by 2016. More information on the historical operations of the Colorado River, as well as information on operations for the upcoming year, can be found in the Colorado River Annual Operating Plans.[70]

Below is a brief summary of a case study investigating the economic impacts of restricted operations at the Glen Canyon Dam under an operational strategy called the Modified Low Fluctuating Flows (MLFF) Alternative.[71]

The MLFF, which was designed to minimize negative impacts on the downstream environment, was developed as a result of a multi-year EIS process initiated in early 1990. As approved by the record of decision on October 8, 1996, it is the current operating regime for the Glen Canyon Dam.

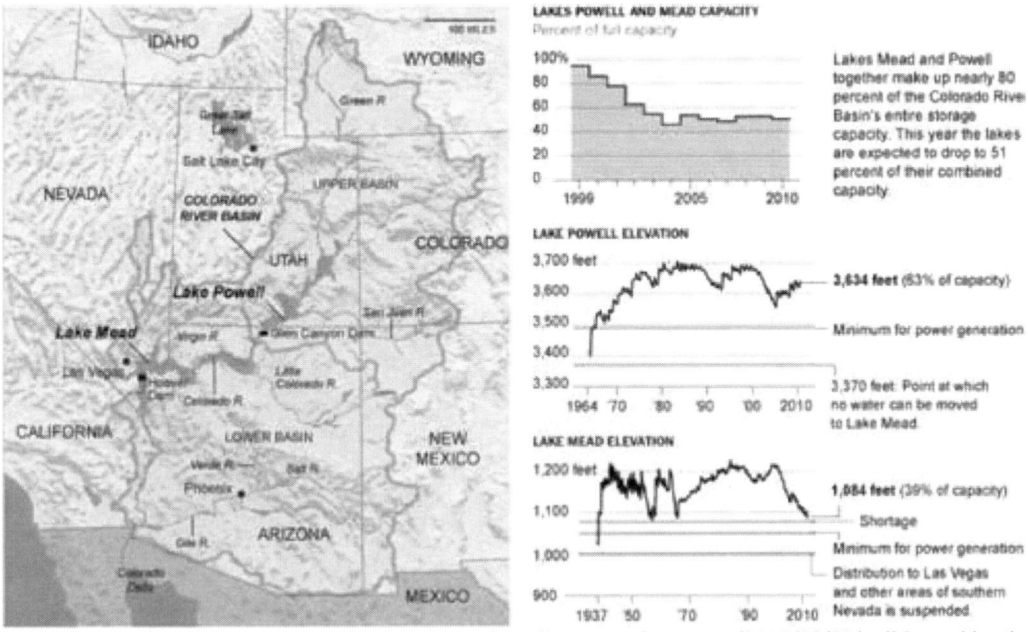

Source: New York Times, September 27, 2010, http://www.nytimes.com/2010/09/28/us/28mead.html, accessed January 4, 2011.

Figure 15. Lakes Powell and Mead.

Economic Impacts of Restricted Operations at Glen Canyon Dam

Glen Canyon Dam is a large hydropower facility (see table 1) on the Colorado River in Arizona, designed and operated historically to produce power primarily during on-peak periods when it is most valuable.

However, the production of peaking power resulted in large fluctuations in downstream releases that caused considerable adverse impacts on the downstream environment. To mitigate these impacts, DOI initiated a new operational strategy in 1996 called the Modified Low Fluctuating Flows .

The MLFF set new restrictions on maximum and minimum flows, ramp rates, and the daily change in flow, with the goal of protecting downstream resources while allowing efficient power production. The new restrictions reduced the generating capacity and limited the ability of the hydropower plant to respond to changes in load in such a way that less energy is generated during the on-peak hours and more energy is generated during the off-peak hours when it is less valuable. A study by Argonne National Laboratory evaluated power economic impacts and compared the results to the economic analysis performed prior to the MLFF. It estimated the annual economic loss resulting from the MLFF implementation to range from approximately $15.1 million to $44.2 million in terms of 1991 dollars ($1991).

Source: Ex Post Power Economic Analysis of Record of Decision Operational Restrictions at Glen Canyon Dam, Argonne National Laboratory, July 2010.

3.3. The Tennessee River System

The Tennessee River system territory includes most of Tennessee and parts of Alabama, Georgia, Kentucky, Mississippi, North Carolina, and Virginia, serving more than 8.7 million people.[72] TVA manages the Tennessee River and its reservoirs as a whole, regulating the flow of water through the river system for flood control, navigation, power generation, water quality, and recreation. TVA is also the Nation‘s largest public power provider, wholly owned by the U.S. Government; it maintains 29 conventional hydroelectric dams (see figure 16).

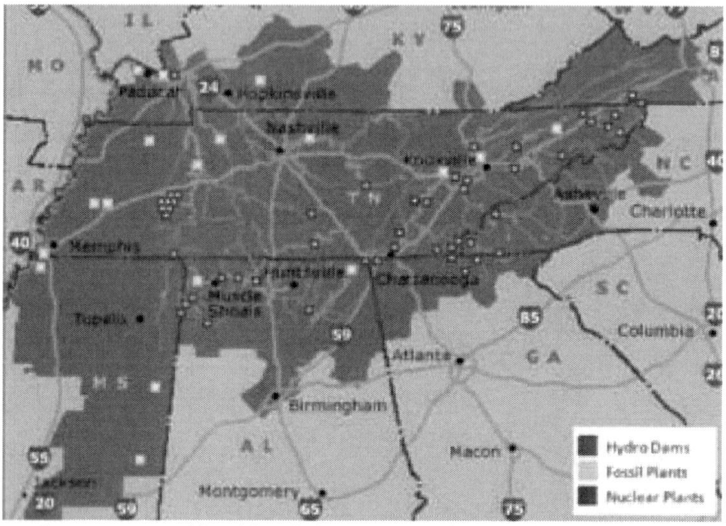

Source: TVA, http://www.tva.com/sites/sites ie.htm, accessed December 12, 2010.

Figure 16. Tennessee Reservoir and Power Plants.

Reservoir-specific flow requirements keep the riverbed below that reservoir‘s dam from drying out, whereas system-wide flow requirements ensure that enough water flows through the river system to meet downstream needs. To meet these requirements, TVA constantly monitors various factors affecting the reservoir inflows, including precipitation and weather patterns, to forecast river conditions to ensure adequate preparation and planning (see figure 17 for the 2010 TVA operating curves). On average, the Tennessee Valley gets 51 inches of rain a year, which is more than double the average rainfall in the southwestern United States.[73]

Nonetheless, the Tennessee V al ley has experienced water shortages during the 2007-2008 droughts that forced communities around the watershed to restrict water withdrawals and take conservation measures. In December 2010, Gary Springston, TVA program manager for water supply, stated that the present situation was still tenuous and —even systems connected to the Tennessee River system could face conflicts between instream flow needs to support water quality and aquatic life and withdrawals for offstream uses such as public-water supply, industry, thermoelectric power generation, and irrigation."[74] Water supply concerns continue to increase due to population growth and interbasin transfers, especially since the Tennessee River is surrounded by areas that may require more water to accommodate growing needs.

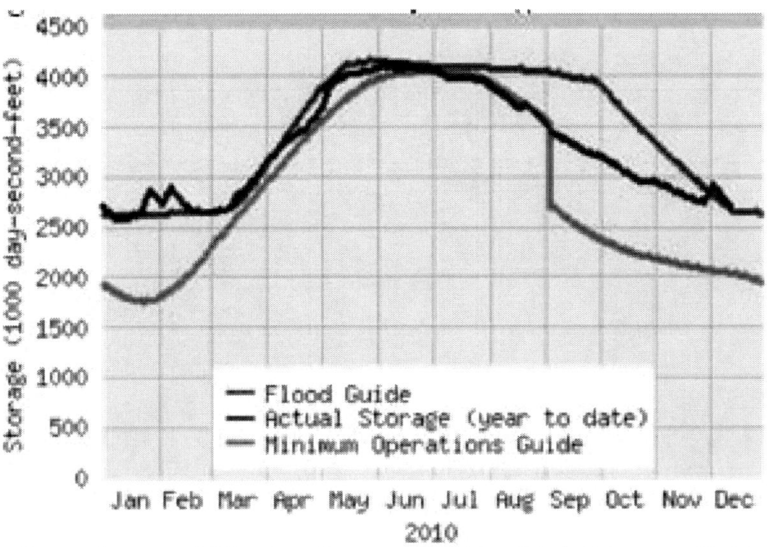

Source: TVA River System, TVA, http://www.tva.gov/river/lakeinfo/systemwide.htm, accessed December 12, 2010.

Figure 17. Tennessee River Operating Curves.

3.2.1. Effects of Droughts on Hydroelectric Power Generation

The 2007-2008 droughts in the TVA region were among the worst on record, during which low reservoir water levels caused TVA to lose almost half of its total hydroelectric generation[75] (see figure 18). At the same time, coal prices more than doubled, forcing TVA to rely on additional natural gas purchases to meet electric generation needs while keeping prices as low as possible. Even with the increased reliance on natural gas as opposed to coal, TVA raised rates by 20 percent in October 2008 to absorb more than $2 billion of increased costs for coal, natural gas, and purchased power[76] (see figure 19).

To address these and other water supply issues, the Tennessee Valley Water Partnership Drought Committee was formed in 2004. The committee consists of representatives from the following States and agencies: TVA, Alabama, Georgia, Kentucky, Mississippi, North Carolina, Tennessee, Virginia, the U.S. Environmental Protection Agency, and the U.S. Geological Survey.[77]

The committee is activated when drought conditions are severe or worse with tributary reservoir levels below the system minimum operating guidelines, and serves as a forum for the exchange of information and views by the participants.

The Tennessee River Basin, like other major watersheds, crosses State lines. As noted in the December 2010 Tennessee River and Reservoir system update from TVA, watersheds are natural —systems that [do not] follow manmade jurisdictions, and effective water supply planning has to reflect this reality.

For example, drought conditions may exist throughout a particular watershed that sits astride a State line. One State has imposed water conservation measures, while the other has not. In the absence of a coordinated response to drought conditions, people living in the _downstream' State may experience water shortages. Water is a shared resource; [therefore, it is] important to plan from a regional perspective.[78]

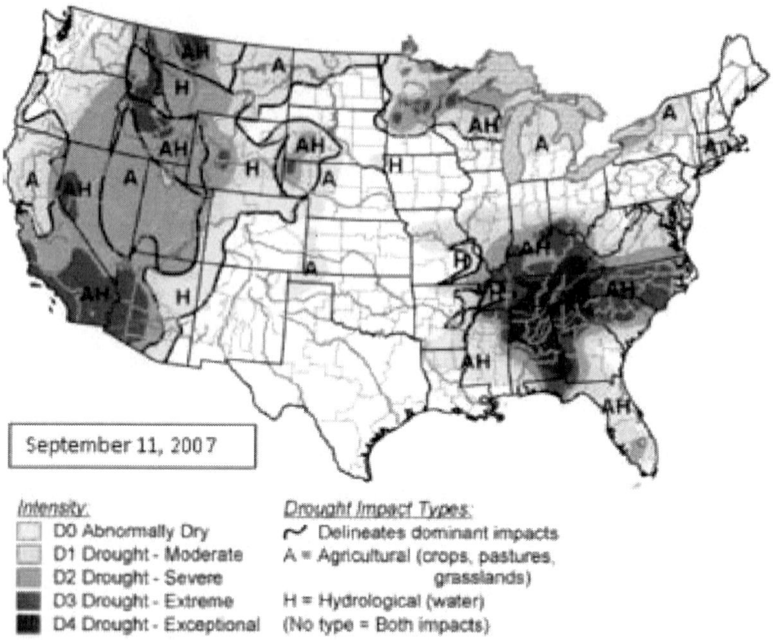

Source: U.S. Drought Monitor Archives, http://www.drought.unl.edu/dm/archive.html, accessed January 9, 2011.

Figure 18. Drought Observed in 2007.

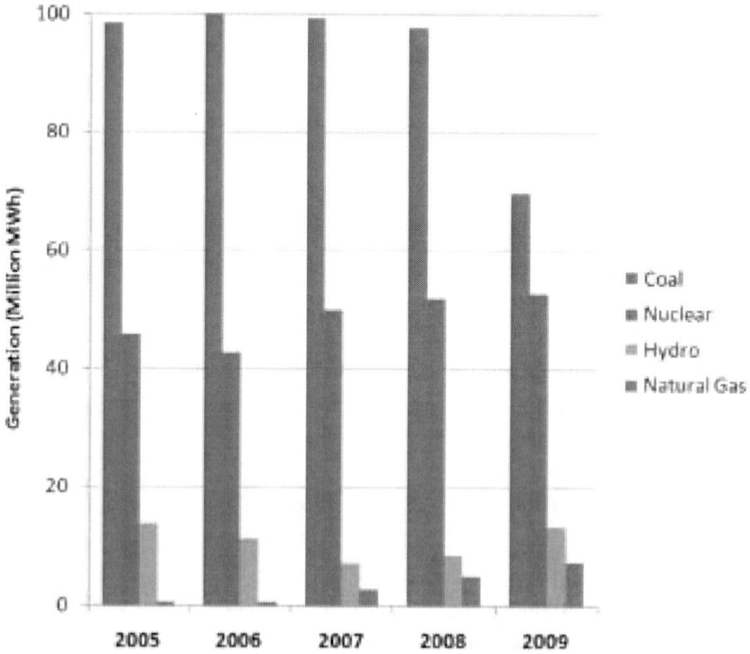

Source: Derived from EIA-906, EIA-920, EIA-923 databases, http://www.eia.doe.gov/cneaf/electricity/page/eia906 920.html, accessed November 30, 2010.

Figure 19. TVA Generation Profile, 2005-2009.

SECTION 4: DISCUSSIONS WITH HYDROELECTRIC FACILITY OWNERS AND OPERATORS

The insights of owners and operators of hydroelectric facilities are essential in understanding not only the day-to-day operation of critical hydropower infrastructure, but also the key issues that the owners and operators must consider in their long-term planning. For that reason, DOE and DHS engaged in discussions with senior hydroelectric facilities personnel who volunteered to participate in this study.

The primary objective of the discussions was consistent with the purpose of the study as described in section 1 – to examine the issues related to the effects of weather pattern variability on the overall management of reservoirs and streamflows at dams. Discussion participants were experienced operations personnel representing public and private owners and operators of hydro conventional, pumped storage, and run-of-the-river hydroelectric plants. A hydropower consumer organization representative also participated in the discussion. Owners and operators from all geographic regions, with the exception of the Northeast, were represented in the discussions.

There was substantial agreement among discussion participants that they were generally able to produce power even in low water situations and that high water events could be more problematic than low water events. They all agreed that the most significant issue facing them in producing hydropower is competing demands for the use of the available water.

The storage capacity and conveyance flexibility of most impoundment facilities were designed to accommodate local or regional historical patterns of hydrologic variability. Thus, episodic low water conditions, as opposed to long-term drought conditions that could and have affected hydropower production, are not critical contributors to reduced hydropower production; scientific and technical advances can limit their effect. These same advances, however, do not address the requirements for providing sufficient water for irrigation, environmental protection, community and industrial uses, or transportation that in many cases are already in conflict. Low water or drought conditions resulting from weather pattern variability will only heighten the degree of competition for available water.

4.1. Low Water Condition Accommodations

The science of accommodating hydropower production to low water conditions lies in the increasingly sophisticated modeling that allows optimization of the water that is available to produce power. An owner of several dams indicated that the facilities are able to operate at low water levels and they benefit if there is more water available for use than expected. Technological advances can further optimize the use of available water through modernization.

Several operators reported that more generation can be achieved from lower head dams through turbine upgrades. Such investment can realize a three to five year payback period as a result of efficiency gains. An operator of hydroelectric facilities reported an increase of 560 MW in overall hydro capacity through turbine upgrades at several facilities; another indicated that it received tax credits for upgrades qualified under certain Federal energy efficiency programs. Pumped storage units are another technological approach to withstand the possibility of reduced precipitation. In pumped storage facilities, water stored in a higher elevation reservoir is released through turbines to a lower elevation reservoir to produce power during high demand periods. The water is then pumped back up to the higher elevation reservoir using low-cost, off-peak electric power. Pumped storage facilities are more resilient to unexpected weather pattern changes, including extreme low water events, because the water used to produce power is stored (in the reservoirs) and recycled (i.e., not released into the natural stream flow).

The discussion participants shared another way to cope with low water conditions – hydroelectric power infrastructure modernization. However, they also indicated that costs associated with infrastructure modernization can become an issue. Financial resources to design and implement facility upgrades generally come through public funds and/or power sales for publicly held hydropower infrastructure, and from rate increases approved by public utility commissions for privately held facilities. Although payback periods could be as short as three to five years for technology upgrades, as noted earlier, securing the initial investment can be challenging.Some owners have received offers from investors and other utility companies to enter into a variety of energy savings performance contracts that would provide the initial investment for modernization in return for a share of the subsequent increased energy production. None of the participants indicated that they were presently involved in such contracts and several raised concerns as to whether they could legally enter into such arrangements. The potential for technology upgrades at some hydropower infrastructure may also be limited or made more expensive due to the age or physical condition of the facility.

4.2. High Water Condition Impacts

High water conditions present a unique set of challenges that are more problematic than low water conditions. Although operators want to retain as much water as possible in the reservoir for hydropower production, storing it in the reservoir during high water conditions may be hard to manage, as it might impact residences surrounding the reservoir.

Many dams have multiple missions; for some, the requirement for flood control takes precedence over hydropower production. Adherence to this primary mission may require passing high volumes of water through the dam turbines even though there may be low power demand.

These increased flows may also require downstream dams to pass through water and not be able to sell the resulting power at a reasonable price. Even if flood control is not a facility mission, owners do their best to avoid or minimize downstream harm when they manage high water conditions. Debris buildup associated with flooding can be dangerous to the facility infrastructure and affect operations. Trees, lumber, sheds, animals, and other debris can be swept into rivers from floods and can build up against dams. The cost and personnel resources required to remove this debris can be significant.

4.3. Competing Demands for Water

Hydroelectric facilities serve multiple purposes that can include flood control, recreation, industrial and community water supply, irrigation, and transportation. The demands for water for these uses can come into conflict with hydropower production in terms of how much water can be used for nonpower generation and the condition of the water associated with power generation.

For multifunction facilities, the combination of existing water rights, treaties, contracts, laws, or court cases determine who gets how much water and when they receive it. Modifying these controlling forces to consider reduced water availability can be difficult because they may involve multiple States and parties, and sometimes, international partners. In addition to these legally binding obligations on water delivery, softer forces, such as providing or storing water to protect recreational uses or the value of residences around the reservoir, can also limit the availability of water for hydropower generation.

The condition of the water used in producing hydropower may also be heavily controlled through Federal and State laws and regulations, operating permits and licenses, and court cases related to the protection of natural resources and the environment. These controlling forces may stipulate water conditions such as tail water temperature, streamflow, and dissolved oxygen levels.

Operating stipulations are primarily designed to protect species designated as threatened or endangered under Federal or State laws. They may also serve to protect downstream banks, channels, and river branches.

All discussion participants voiced their support for natural resources and environmental protection; however, they also expressed concerns that hydropower production interests are not always evenly represented when these environmental protection requirements are developed. Several participants described examples of how compliance with these operating restrictions limits optimal operation of hydropower facilities.

A hydropower operator indicated that certain facilities could not operate in certain months in order to maintain the required tailwater temperature. Another noted that complying with the required dissolved oxygen levels applicable to the facilities is decreasing turbine efficiency thereby affecting facility performance. Requirements in one operator‘s permit specify minimum flow schedules that are higher than the historical minimum flow. Maintaining the stipulated streamflow requires generating power when there is no or low demand. Operators reported that in cases such as these, a hydropower producer could be forced to incur costs to produce power.

Competing demands for water are already evidenced in several parts of the country.[79] The operational constraints associated with natural resource and environmental protection will only increase in their intensity if projected climate change results in increasingly unpredictable water availability.

APPENDIX A. ACRONYMS

BPA	Bonneville Power Administration
CIPAC	Critical Infrastructure Partnership Advisory Council
DHS	U.S. Department of Homeland Security
DOE	U.S. Department of Energy
DOI	Department of Interior
EIA	U.S. Energy Information Administration

MLFF	Modified Low Fluctuating Flows
MW	Megawatt
MWh	Megawatt hours
NIPP	National Infrastructure Protection Plan
NOAA	National Oceanic and Atmospheric Administration
PNCA	Pacific Northwest Coordination Agreement
SPI	Standardized Precipitation Index
SSA	Sector-Specific Agency
TVA	Tennessee Valley Authority
USACE	U.S. Army Corps of Engineers
USGS	U.S. Geological Services

APPENDIX B. GLOSSARY OF TERMS

Black start Capability: refers to the ability of a generating unit or station to start operating and delivering electric power without assistance from the electric system. Black start units are essential to restart generation and restore power to the grid in the event of an outage.

Critical rule curve: defines the reservoir elevations that must be maintained to ensure that firm hydro energy requirements can be met under the most adverse streamflows on record.

Drought: is a period of unusually persistent dry weather that persists long enough to cause serious problems such as crop damage and/or water supply shortages. The severity of the drought depends upon the degree of moisture deficiency, the duration, and the size of the affected area.

Drought can be defined in four different ways:

- Meteorological: a measure of departure of precipitation from normal. Due to climatic differences, what might be considered a drought in one location of the country may not be a drought in another location;
- Agricultural: refers to a situation where the amount of moisture in the soil no longer meets the needs of a particular crop;
- Hydrological: occurs when surface and subsurface water supplies are below normal; and
- Socioeconomic: refers to the situation that occurs when physical water shortages begin to affect people.[80]

Flood control curve: defines the drawdown required to ensure adequate space is available in the reservoir to regulate the predicted runoff for the year without causing flooding downstream.

Peaking capacity: capacity of generating equipment normally reserved for operation during the hours of highest daily, weekly, or seasonal loads. Some generating equipment may be operated at certain times as peaking capacity and at other times to serve loads on an around-the-clock basis.

Renewable Energy Credits (RECs): represents the property rights to the environmental, social, and other nonpower qualities of renewable electricity generation. A REC, and its associated attributes and benefits, can be sold separately from the underlying physical electricity associated with a renewable-based generation source.

Streamflow: the rate and volume of water flowing in various sections of a river.

Watershed: is the area of land that drains to a particular water body. A watershed is defined by the highest elevations surrounding a water body, consistent with the concept that —water runs downhill."

APPENDIX C. STUDY METHODOLOGY

DOE initiated this study in collaboration with DHS under the Critical Infrastructure Partnership Advisory Council (CIPAC) framework.[81] Under the CIPAC, critical infrastructure stakeholders in government and private sectors formed a partnership model and a forum in which they can engage in a broad spectrum of activities to support and coordinate critical infrastructure protection, security, and resilience.[82]

In addition to reviewing the 2009 Sandia report, DOE conducted exhaustive Internet and literature research on several related topics including drought, precipitation, weather changes, streamflow regulation, water management and uses, as well as the operation of hydroelectric facilities (see appendix F).

DOE also examined EIA's annual electricity survey forms EIA-906, EIA-920, EIA-923, and EIA-860 to investigate the historical pattern of electric power generation and generating capacity.

The focus of the analysis is on hydroelectric power generation at the State level and the plant level in the last 20 years from 1990 to 2009.

The operation of, and planning at, hydroelectric facilities are driven by a number of factors that are often unique to each watershed; therefore, DOE assessed several watersheds to help better understand the function of hydroelectric dams in the context of a large river system.

Based on the amount of hydropower generation and the gravity of water-related concerns affecting hydropower, the following three rivers were selected as an example: the Columbia River, the Colorado River, and the Tennessee River. The reservoirs in these rivers serve multiple purposes, many of which often conflict with one another, including public water supply, irrigation, flood control, fish habitat protection, and power generation. DOE considered these issues that are often heightened during low water periods when the availability of water diminishes.

Following a literature review, DOE and DHS engaged a small number of Dams Sector owners and operators, both public and private, to seek their expertise and insight. The goal of the discussions was to ascertain the most critical issues they are faced with today and in the near future in their operation of hydroelectric facilities.

APPENDIX D. IMPACTS OF DROUGHTS ON UTILITIES

Company Name & Year	Impacts of Drought[83]
BC Hydro[84] 2010	"Severe drought conditions in northeast British Columbia have BC Hydro bracing for a $220 million increase in electricity importsthis fiscal year… because of low water volumes coming into its hydroelectric reservoirs." –July 2010, Vancouver Sun
Southern Co.[85] 2007	"Georgia Power's hydroelectric power generation was down 51 percent in 2007, forcing the company to spend $33.3 million for purchasing coal and oil to replace lost hydropower generation although hydropower sources account for less than twopercent of Georgia Power's generation portfolio." –Nov. 2007, Atlanta Business Chronicles
Manitoba Hydro[86] 2003	"A net loss of $436 million was reported in Manitoba Hydro's 53rd annual report for the fiscal year ending March 31, 2004.The loss was primarily due to the prolonged drought conditions that affected normal electricity production at the utility's 14 hydroelectric generating stations." –2004, Manitoba Hydro
WAPA[87] 2005	"Continued drought across the West has caused lower water inflows, which, in turn, create decreased reservoir storage levels and decreased available capacity and energy. When generation is not sufficient to meet firm power contract commitments, Western will purchase power from other suppliers…Western spent almost $498 million in FY 2005 Western-wide for 11.7 million MWh of purchased power, compared to almost $404 million in FY 2004 for 13.2 million MWh." –About WAPA: F&Q
WA State[88] 2005	"Looking at the Statewide financial impact of the drought on electricity costs, our analysis indicates that the 2005 drought could potentially increase the cost of supplying electricity by 199 to 313 million dollars.This would be the equivalent of 4-7% overall increase in electricity costs to Washington consumers." –May 2005, WA State Dept. of Community

APPENDIX E. 2009 SANDIA REPORT

The work conducted in this report builds upon prior energy and water research conducted by Sandia National Laboratories – New Mexico and issued in a report titled, *Energy and Water Sector Policy Strategies for Drought Mitigation*. The following summary of that report focuses on the use of water for fossil-fuel fired power generation – cooling in thermoelectric power plants and in movement of coal by barge. Unless otherwise noted, information and data presented in the section below are taken from the 2009 Sandia report.

Water Use in Cooling Thermoelectric Power Plants

Water is used as the primary coolant in the condensers in both steam and natural gas-fired, combined cycle plants; the amount of water used for cooling in these plants can be significant, depending on the type of cooling system used. Plants that use "once-through" or "open-loop" cooling systems withdraw large amounts of water from nearby surface water sources. This water passes through a condenser as a coolant and, in doing so, transfers heat energy from the hot steam to the coolant water, raising the temperature of the water. After moving through the condenser, the water is released to the original lake, pond, or river source. The increased temperature of the discharge water also increases the rate of evaporation for the body of water. The quantity of water lost from the hydrological system by evaporation caused by elevated temperatures is said to be —consumed.

Closed-loop cooling systems withdraw smaller volumes of water than open-loop systems because they involve a mechanism for circulating a portion of the coolant water.[89] Other closed-loop systems use cooling towers. In wet cooling towers, hot water that is discharged from the condenser is sprayed over metal plates while a fan blows cool air up the tower. The water that evaporates is considered to be consumed, while the remaining water falls down the tower and can be returned to the condenser. Dry cooling towers pump the hot water in small pipes down the tower as fans blow cool air over the pipes. The cooled water is returned to the condenser. This cooling approach is considerably more expensive to operate and is inefficient in hot, arid climates.

Water use and consumption data indicate that fossil and nuclear steam turbines with open-loop systems operate with high water use intensity.[90] In 2005, power plants equipped with once-through cooling systems accounted for 92 percent of water withdrawals for thermoelectric power while plants equipped with re-circulating systems withdrew the remaining 8 percent.[91] Cooling technologies that require less water and allow for the production of thermoelectric power are common in areas where water is scarce or strictly managed such as Nevada, New Mexico, and Utah.

Coal Transport by Barge

Transportation on the inland waterways and Great Lakes is an important element of the domestic coal distribution system, carrying approximately 20 percent of the Nations‘ coal, enough to produce 10 percent of U.S. electricity annually.[92] Barge transport is often used to transfer coal from the initial source to a railroad, from a railroad to the coal-fired power plant, or the entire distance from the mine to the plant. Barge traffic is particularly important in the Midwestern and Eastern States, with 80 percent of shipments originating in States along the Ohio River. The amount of waterborne transported coal has remained relatively constant over the last two decades.[93]

Barge transport and the amount transported on a single barge are dependent upon the depth of the river on which the barge travels. Reducing the barge load is costly. Losing one foot of draft typically means losing 17 tons of cargo on a single barge and 255 tons on a typical 15-barge tow. In addition, idle tow-boats cost shipping companies $5,000 - $10,000 per day.[94] Droughts have the potential to reduce the rate at which all goods, including coal, can be transported by barge.

Some river systems, like the Missouri River, have a system of reservoirs that are used to control river depths. When river levels are low, water is released from the reservoirs to increase river depths and permit barge travel. To mitigate the potential for low water levels to significantly disrupt electric power generation, most coal-burning plants with barge access can also receive coal shipments by rail. However, because barge is the cheapest mode of transportation, utilities pay a higher rate for transportation.

Environmental Concerns

Thermoelectric power water withdrawals have been affected by limited water availability in some areas of the United States and also by sections of the Clean Water Act that regulate cooling system thermal discharges and mandate the best use of available technology for minimizing environmental effects of cooling water intake. Consequently, since the 1970s, power plants have increasingly been built with, or converted to, cooling towers, ponds, or dry re-circulating systems instead of once-through cooling systems.[95]

By affecting the availability of cooling water, drought has had an impact on the production of electricity from thermoelectric power plants. The problem for power plants becomes acute when river, lake, or reservoir water levels fall near or below the level of the water intakes used for drawing water for cooling. A related problem occurs when the temperature of the surface water increases to the point where the water can no longer be used for cooling. The Southeast experienced particularly acute drought conditions in August 2007, which forced the shutdown of some nuclear power plants and curtailed operations at others in order to avoid exceeding environmental limits for water temperature. A similar situation occurred in August 2006 along the Mississippi River, as well as at some plants in Illinois and Minnesota.[96]

Changes in Utility Water Consumption over Time

The United States Geological Service's (USGS) National Water Use Information program prepares nationwide compilations of all reported water uses that are published every five years.[97] According to the USGS, more than 340,000 million gallons of water is withdrawn per day in the United States, 30 percent of which is "consumed" and 70 percent is returned to flow. [98]

The Reclamation-led effort resulted in the development of the *Colorado River Interim Guidelines for Lower Basin Shortages and Coordinated Operations for Lake Powell and Lake Mead* (Interim Guidelines) that sought to address the unique challenges in the operation of the Colorado River. 99 This effort included an environmental review and public scoping meetings. Thermoelectric freshwater withdrawals accounted for 41 percent of all freshwater withdrawals in 2005; however, it is important to note that only 3 percent of the withdrawn water is consumed and the rest is returned to natural flow.

Nearly all of the water withdrawn was surface water used for once-through cooling at power plants. Twenty-nine percent of thermoelectric power withdrawals were from saline or brackish coastal water bodies.

Thermoelectric power withdrawals in 2005 were 3 percent more than in 2000 due to an increase in demand for electricity. USGS data indicates that water use for thermoelectric power increased from 1950 to 1980, but has remained relatively stable from then through 2005.[100] As noted above, the 1980 Federal Clean Water Act requirements resulted in an end to construction of open-loop cooling systems for cooling thermoelectric power plants and use of closed-loop cooling. Closed-loop systems require much lower consumption of surface water.

Hydroelectric facilities use water as the source of energy that is converted to electric power. Although the operation of hydroelectric facilities depends on water flow, the water consumption by hydro facilities is negligible.

APPENDIX F. SOURCES AND REFERENCES

—1997 Pacific Northwest Coordination Agreement," U.S. Army Corps of Engineers, June 18, 1997, http://www.nwd-wc.usace.army.mil/PB/oper planning/97PNCA Conformeipdf.

American Waterway Operators, www.americanwaterways.com/industry stats/index.html.

—Apalachicola-Chattahoochee-Flint (ACF) Drought: Federal Reservoir and Species Management," Congressional Research Service, RL34250, Updated May 1, 2008, http://assets 20071114.pdf.

Bates, B.C., Z.W. Kundzewicz, S. Wu, and J.P. Palutikof (eds.), 2008: Climate Change and Water. Technical paper of the Intergovernmental Panel on Climate Change. IPCC Secretariat, Geneva, Switzerland, available at http://www.ipcc.ch/pdf/technical-papers/climate-

BPA. —The Columbia River System: Inside Story," Second Edition, Bonneville Power Administration, April 2001, http://www.bpa.gov/power/pg/columbia river inside story.pdf.

BPA. 2009 BPA Facts, Bonneville Power Administration, http://www.bpa.gov/corporate/ about BPA/Facts/FactDocs/2009 BPA Facts.pdf.

BPA. 2010 Annual Report, Bonneville Power Administration, http://www.bpa.gov/ corporate/ finance/a report/10/AR2010.pdf.

BPA. BPA's Interim Environmental Redispatch and Negative Pricing Policy, Administrator's Final Record of Decision, May 2011, Bonneville Power Administration, http://www.bpa.gov/corporate/pubs/RODS/2011/ERandNegativePricingFinalROD web.pdf.

BPA. Columbia River high water management, Bonneville Power Administration, http://www.bpa.gov/corporate/AgencyTopics/ColumbiaRiverHighWaterMgmnt/.

BPA. Columbia River high-water operations, [June 1-14, 2010]," Bonneville Power Administration, September 2010, http://www.bpa.gov/corporate/pubs/final-report-columbiariver-high-water-operations

BPA. Federal Columbia River Power System, Bonneville Power Administration, http://www.bpa.gov/power/pgf/hydrpnw.shtml.

BPA. Statement on Environmental Redispatch and Negative Pricing, Bonneville Power Administration, December 3, 2010, http://www.bpa.gov/corporate/Agency Topics/ ColumbiaRiverHighWaterMgmnt/Environmental %20Redispatch%20statement.pdf.

Bull, S. R., D. E. Bilello, J. Ekmann, M. J. Sale, and D. K. Schmalzer, 2007: Effects of climate change on energy production and distribution in the United States in Effects of Climate Change on Energy Production and Use in the United States. A Report by the U.S. Climate Change Science Program and the Subcommittee on Global Change Research. Washington, DC.

Coal: Research and Development to Support National Energy Policy, Committee on Coal Research, Technology, and Resource Assessment to Inform Energy Policy, National Research Council, 2007.

"Colorado River Basin Water Management: Evaluating and Adjusting to Hydroclimatic Variability," Committee on the Scientific Bases of Colorado River Basin Water Management, National Research Council, 2007, http://www.nap.edu/catalog/11857.html.

Colorado River Compacts, The Colorado River Water Conservation District, April 16, 2008, http://www.crwcd.org/media/uploads/20080416 CO River Compact.pdf.

Colorado River Law and Policy, Western Water Assessment, University of Colorado – Boulder, http://wwa.colorado.edu/colorado river/law.html.

Database of State Incentives for Renewables & Efficiency, June 2011, http://www.dsireusa.org/documents/summarymaps/RPS map.pptx.

Dave Flessner, "Tennessee Valley Authority boosts rates 20 percent," Chattanooga Times Free Press, August 21, 2008, http://www.timesfreepress.com/ news/2008/aug/21/ tennessee-valleyauthority-boosts-rates-20-percent/?local.

DHS. 2010 Dams Sector-Specific Plan, U.S. Department of Homeland Security, http://www.dhs.gov/xlibrary/assets

DHS. Critical Infrastructure Partnership Advisory Council, DHS http://www.dhs.gov/files/ committees/editorial 0843.shtm.

DHS. National Infrastructure Protection Plan, U.S. Department of Homeland Security, http://www.dhs.gov/xlibrary/assets Plan.pdf.

DOE. "Feasibility Assessment of the Water Energy Resources of the United States for New Low Power and Small Hydro Classes of Hydroelectric Plants," U.S. Department of Energy, Office of Energy Efficiency and Renewable Energy, January 2006, http://www1.eere.energy.gov/windandhydro/pdfs/doewater-11263.pdf.

DOE. 2010 Energy Sector-Specific Plan, U.S. Department of Energy, http://www.oe.energy SSP 2010.pdf.

DOE. Energy Demands on Water Resources, Report to Congress on the Interdependency of Water and Energy, U.S. Department of Energy, December 2006.

DOE/NETL. Impact of Drought on U.S. Steam Electric Power Plant Cooling Water Intakes and Related Water Resource Issues, National Energy Technology Laboratory, DOE/NETL2009/1364, April 2009, http://www.netl.doe.gov/technologies/ coalpower/ ewr/water/

DOE/NETL. "An Analysis of the Effects of Drought Conditions on Electric Power Generation in the Western United States," National Energy Technology Laboratory, DOE/NETL-2009/1365, April 2009, http://www.netl.doe.gov/technologies/ coalpower/ ewr/water/pdfs/final%20- %20WECC%20drought%20analysis.pdf.

DOI. "Secretary Kempthorne Signs Historic Decision For New Colorado River Management Strategies," U.S. Department of Interior, December 13, 2007, http://www.doi.gov/ archive/news/07 News Releases/071213.html.

Drought Management Plan, Tennessee Department of Environment and Conservation, February 2010, http://www.tn.gov/environment

EIA. Annual Energy Outlook 2010, U.S. Energy Information Administration, December 2010, http://www.eia.gov/oiaf/aeo/tablebrowser/#release=AEO2011&subject=10- AE O2011&table=16-AEO2011&region=0-0&cases=ref2011-d120810c.

EIA. EIA-906, EIA-920, EIA-923, and EIA-860 databases, annual electric power generation, U.S. Energy Information Administration, http://www.eia.doe.gov/cneaf/electricity 920.html.

EIA. Electricity Net Generation from Renewable Energy by Energy Use Sector and Energy Source, U.S. Energy Information Administration, http://www.eia.doe.gov/cneaf/ solar.renewables/page/table3.html.

EIA. Glossary, U.S. Energy Information Administration, http://www.eia.doe.gov/glossary/index.cfm.

EIA. Net Generation by Energy Source, U.S. Energy Information Administration, http://www.eia.doe.gov/cneaf/electricity 1.html.

Electricity Impacts of the Drought, State of Washington, Department of Community, Trade, and Economic Development, May 9, 2005, http://www.commerce.wa.gov/ cted/documents/ID 2015 Publications.pdf.

Eric Wolff, "ENERGY: Hoover Dam could stop generating electricity as soon as 2013, official fear," North County Times, September 11, 2010, http://www.nctimes.com/business/article b7e44e9e-087d-53b2-9c49-7ea32262c9a9.html.

"Ex Post Power Economic Analysis of Record of Decision Operational Restrictions at Glen Canyon Dam," Argonne National Laboratory, July 2010, http://www.wapa.gov/CRSP/environment/documents/Post-RODFinal 000.pdf.

Federal Water Projects on the Colorado River System, PBS, http://www.pbs.org/cowboys indianslawyers/topicfeature.html.

Felicity Barringer, "Water Use in Southwest Heads for a Day of Reckoning," New York Times, September 27, 2010, http://www.nytimes.com/2010/09/28/us/28mead.html.

Global Climate Change Impacts in the United States, Thomas R. Karl, Jerry M. Melillo, and Thomas C. Peterson, (eds.). Cambridge University Press, 2009, http://downloads.globalchange.gov/usimpacts/pdfs/climate-impacts-report.pdf.

Gutowski, W.J., G.C. Hegerl, G.J. Holland, T.R. Knutson, L.O. Mearns, R.J. Stouffer, P.J. Webster, M.F. Wehner, and F.W. Zwiers, 2008: Causes of observed changes in extremes and projections of future changes. In: Weather and Climate Extremes in a Changing Climate: Regions of Focus: North America, Hawaii, Caribbean, and U.S. Pacific Islands [Karl, T.R., G.A. Meehl, C.D. Miller, S.J. Hassol, A.M. Waple, and W.L. Murray (eds.)]. Synthesis and Assessment Product 3.3. U.S. Climate Change Science Program, Washington, DC, pp. 81-116.

Hightower, M., Energy and Water: Overview of Emerging Issues and Challenges, DOE/EIA 2010 Energy Conference.

Hydropower and the World's Energy Future, International Hydropower Association, November 2000, http://www.ieahydro.org/reports/Hydrofut.pdf.

—Hydropower is Reliable," National Hydropower Association, http://hydro.org/why-hydro/reliable/.

Hydropower, the National Energy Education Development Project, http://www.need.org/needpdf/infobook activities/SecInfo/HydroS.pdf.

James Cartledge, —Hoover Dam turbines set for upgrade to cope with drought," Brighterenergy.org, April 19, 2010, http://www.brighterenergy.org/9098/news/marine-hydro/hoover-dam-turbines-set-for-upgrade-to-cope-with-drought/.

Janet Zimmerman, —REGION: Colorado River drought threatens power production," The Press-Enterprise, October 6, 2010, http://www.pe.com/localnews/stories/PE News Local D mead06.2b8c2c6.html.

Jim Mann, the Western News, November 16, 2010, http://www.thewesternnews.com /news/article 3435b546-f1ea-11df-b434- 001cc4c002e0. html?mode=print.

Justin Rubner, —Drought hits Hydropower," Atlanta Business Chronicle, November 16, 2007, http://atlanta.bizjournals.com/atlanta/stories/2007/11/19/story2.html.

Kelic, Andjelka, V. Loose, V. Vargas, and E. Vugrin, 2009, Energy and Water Sector Policy Strategies for Drought Mitigation, Sandia National Laboratories, Albuquerque, NM, http://prod.sandia.gov/techlib/access-control.cgi/2009/091360.pdf.

Kenny, J.F., Barber, N.L., Hutson, S.S., Linsey, K.S., Lovelace, J.K., and Maupin, M.A., 2009, Estimated use of water in the United States in 2005: U.S. Geological Survey Circular 1344, 52 p.; Solley, Wayne B., Pierce, Robert R., Perlman, Howard A., 1998, Estimated use of water in the United States in 1995: U.S. Geological Survey Circular 1200.

Kevin McCullen, —Low snowpack to cost BPA projected $233M," Tri-city Herald, May 4, 2010, http://www.tri-cityherald.com/2010/05/04/1001094/low-snowpack-to-cost-bpa-projected.html.

Lynn Muench, American Waterways Operators, quotes in New York Times, August 15, 2005, http://www.nytimes.com/2005/08/15/national/15drought.html? r=1.

Manitoba Hydro 2002-2004 Drought Risk Management Review, January 15, 2005, http://www.hydro.mb.ca/regulatory affairs/electric/gra 08 09/information requests/ Appendix 43-Report on 2002-2004 Drought.pdf.

Martin-Amouroux, Jean-Marie, the economics of hydroelectricity, Hydro 21, Grenoble European Center on Hydropower, June 2004, http://www.hydro21.org/div media/pdf/pdf economie en.pdf.

NOAA. Columbia River Basin Map, National Oceanic and Atmospheric Administration, http://www.nwr.noaa.gov/Salmon-Hydropower/Columbia-Snake-Basin/upload/Col-Basinmap.pdf.

NOAA. Ingram, H., D. Feldman, N. Mantua, K.L. Jacobs, D. Fort, N. Beller-Simms, and A. M. Waple, 2008: The Changing Context. In: Decision-Support Experiments and Evaluations using Seasonal-to-Interannual Forecasts and Observational Data: A Focus on Water Resources. A Report by the U.S. Climate Change Science Program and the Subcommittee on Global Change Research [Nancy Beller-Simms, Helen Ingram, David Feldman, Nathan Mantua, Katharine L.

Jacobs, and Anne M. Waple (eds.)]. NOAA's National Climatic Data Center, Asheville, NC, pp. 7-28.

NOAA. Western Area Power Administration, Frequently Asked Questions: Power Marketing, http://www.wapa.gov/about/faqpm.htm.

NOAA. "What is Drought?", National Oceanic and Atmospheric Administration, http://www.wrh.noaa.gov/fgz/science

Pontius, Dale, "Colorado River Basin Study: Final Report," The Western Water Policy Review Advisory Commission, August 1997, http://wwa.colorado.edu/colorado river/docs/pontius%20colorado.pdf.

Reclamation. "Addressing Climate Change in Long-Term Water Resources Planning and Management," U.S. Army Corps of Engineers and Bureau of Reclamation, January 2011, http://www.usbr.gov/climate/userneeds/.

Reclamation. "Colorado River Interim Guidelines for Lower Basin Shortages and Coordinated Operations for Lakes Powell and Mead," Bureau of Reclamation, December 2007, http://www.usbr.gov/lc/region/programs/strategies/about.html.

Reclamation. Annual Operating Plans, Upper Colorado Region, Bureau of Reclamation, http://www.usbr.gov/uc/water

Reclamation. "Colorado River Basin Water Supply and Demand Study Plan of Study," Bureau of Reclamation, http://www.usbr.gov/lc/region/programs/crbstudy.html.

Reclamation. Grand Culee Dam, Bureau of Reclamation, Pacific Northwest Region, Bureau of Reclamation, http://www.usbr.gov/pn/grandcoulee/index.html.

Reclamation. Hydroelectric Power, Bureau of Reclamation, July 2005, http://www.usbr.gov/power/edu/pamphlet.pdf.

Reclamation. The Law of the River, Bureau of Reclamation, http://www.usbr.gov/lc/region/g1000/lawofrvr.html.

Reclamation. The Record of Decision, Colorado River Interim Guidelines for Lower Basin Shortages and Coordinated Operations for Lakes Powell and Mead, Bureau of Reclamation, http://www.usbr.gov/lc/region/programs/strategies/RecordofDecision.pdf.

Reclamation. Upper Colorado Region, Bureau of Reclamation, http://www.usbr.gov/uc/ water

"Renewable Portfolio Standards and Renewable Energy Credits," Washington State Legislature, February 10, 2010, http://www.leg.wa.gov/Senate

Scott Simpson, "BC Hydro faces severe drought on Peace River," The Vancouver Herald, July 15, 2010, http://www.vancouversun.com/news/Hydro+faces+severe+drought +Peace+River/3274186/story. html.

SPI Archived Maps, http://www.drought

Stewart, I.T., D.R. Cayan, and M.D. Dettinger, 2004: Changes in snowmelt runoff timing in western North America under a _business as usual' climate change scenario. Climatic Change, http://meteora.ucsd.edu/cap/stewart clch.pdf.

Time-series plot of the annual flow volume (in millions of acre-feet) for the Colorado River at Lee's Ferry, http://wwa.colorado.edu/colorado river/docs/USGS 2004 3062.pdf.

TVA. TVA River System, Tennessee Valley Authority, http://www.tva.gov/river/lakeinfo/systemwide.htm.

TVA. 2007 TVA Strategic Plan, Tennessee Valley Authority, http://www.tva.com/ stratplan/tva strategic plan.pdf.

TVA. Tennessee Reservoir and Power Plants, Tennessee Valley Authority, http://www.tva.com/sites/sites ie.htm.

TVA. Tennessee River and Reservoir System Update, December 6, 2010, http://www.tva.gov/email/eRiver/2010/december.html.

U.S. Census Bureau, http://quickfacts.census.gov/qfd/maps/thematic/PL0120000.html.

U.S. Drought Monitor Archives, National Drought Mitigation Center, http://www.drought

U.S. Geological Survey, Trends in Water Use in the United States, 1950 – 2005. U.S. Geological Survey, USGS Circular 1344, October 2009.

United States Watersheds Map, http://maps.howstuffworks.com/united-states-watersheds-map.htm.

Wind Power: Capacity Factor and Intermittency, Council of European Municipalities and Regions,_http://www.ceere.org/rerl/about wind/RERL Fact Sheet 2a Capacity Factor.pdf.

End Notes

[1] The term —critical infrastructurel has the meaning given to that term in section 1016(e) of the USA PATRIOT Act of 2001. Dams and Energy are two of the original 17 critical infrastructure sectors identified by Homeland Security Presidential Directive 7, which established the United States national policy for identification and

prioritization of critical infrastructure for protection from terrorist attacks. The Critical Manufacturing Sector was added later.

[2] The term —critical infrastructure" has the meaning given to that term in section 1016(e) of the USA PATRIOT Act of 2001. Also see the National Infrastructure Protection Plan, U.S. Department of Homeland Security (DHS), http://www.dhs.gov/xlibrary/assets/NIPP Plan.pdf, accessed December 22, 2010.

[3] See Appendix C for study methodology.

[4] 2010 Dams Sector-Specific Plan, DHS, http://www.dhs.gov/xlibrary/assets accessed December 22, 2010.

[5] 2010 Energy Sector-Specific Plan, U.S Department of Energy (DOE), http://www.oe.energy.gov/ DocumentsandMedia/Energy SSP 2010.pdf, accessed December 22, 2010.

[6] See Appendix B for the meaning of drought and other technical terms used in this report. See Appendix D for examples of effects of drought on utilities.

[7] 2010 Dams Sector-Specific Plan, p. 13. For more information about the possible effects of droughts, see —Apalachicola-Chattahoochee-Flint (ACF) Drought: Federal Reservoir and Species Management," Congressional Research Service, RL34250, Updated May 1, 2008, http://assets.opencrs.com/rpts/RL34250 20071114.pdf and "An Analysis of the Effects of Drought Conditions on Electric Power Generation in the Western United States," National Energy Technology Laboratory, DOE/NETL-2009/1365, April 2009, http://www.netl.doe.gov/technologies/coalpower/ewr/water/pdfs/final%20-%20WECC%20drought% 20analysis.pdf, both accessed January 7, 2011.

[8] Addressing Climate Change in Long-Term Water Resources Planning and Management, the U.S. Army Corp. of Engineers (USACE) and the Bureau of Reclamation (Reclamation), U.S. Department of Interior (DOI), January 2011, http://www.usbr.gov/climate/userneeds/, accessed March 31, 2011.

[9] Throughout this report, "hydroelectric power," "hydropower," and "hydroelectricity" are used interchangeably.

[10] The U.S. Energy Information Administration (EIA), Glossary, http://www.eia.doe.gov/glossary/index.cfm, accessed January 26, 2011.

[11] Kelic, Andjelka, V. Loose, V. Vargas, and E. Vugrin, 2009, Energy and Water Sector Policy Strategies for Drought Mitigation, Sandia National Laboratories, Albuquerque, NM, http://prod.sandia.gov/techlib/access-control.cgi/2009/091360.pdf, accessed December 28, 2010.

[12] Hydroelectric Power, Bureau of Reclamation, July 2005, http://www.usbr.gov/power/edu/pamphlet.pdf, accessed December 21, 2010.

[13] EIA, Net Generation by Energy Source, http://www.eia.doe.gov/cneaf/electricity/epm/table1 1.html, accessed November 30, 2010.

[14] Electricity Net Generation from Renewable Energy by Energy Use Sector and Energy Source, EIA, http://www.eia.doe.gov/cneaf/solar.renewables/page/table3.html, accessed January 7, 2011.

[15] It is important to note that each State treats and defines renewable energy differently. As of March 2011, 29 States have policies in place to provide certain incentives for "eligible renewable sources" (ERS) (see http://www.dsireusa.org/documents/summarymaps/RPS map.pptx). However, electricity generation from large hydropower facilities or those that were operational prior to the implementation of the Renewable Portfolio Standards (RPS) often do not qualify as an ERS. For example, in Washington State, hydro sources generally do not qualify as an ERS, except for incremental electricity produced from efficiency improvements at hydropower facilities owned by qualifying utilities if the improvements were completed after March 31, 1999. See http://www.leg.wa.gov/Senate/Committees/EWE/Documents/RenewableEnergy.pdf, accessed April 4, 2011.

[16] Hydropower, the National Energy Education Development Project, http://www.need.org/needpdf/infobook activities/SecInfo/HydroS.pdf, accessed November 20, 2010.

[17] Hydropower and the World's Energy Future, International Hydropower Association, November 2000, http://www.ieahydro.org/reports/Hydrofut.pdf, accessed May 4, 2011.

[18] "Hydropower is Reliable," National Hydropower Association, http://hydro.org/why-hydro/reliable/, accessed May 4, 2011.

[19] EIA, Form EIA-860 Database Annual Electric Generator Report, 1990-2008, http://www.eia.doe.gov /cneaf/electricity/page/eia860.html, Form EIA-906, EIA-920, and EIA-923 Databases, 1990- 2008, http://www.eia.doe.gov/cneaf/electricity/page/eia906 920.html, both accessed November 12, 2010.

[20] EIA, Annual Energy Outlook 2010, December 2010, http://www.eia.gov/oiaf/aeo/ tablebrowser/# release=AEO2011&subject=10-AEO2011&table=16-AEO2011&region=0-0&cases=ref2011-d120810c, accessed January 25, 2011.

[21] Energy Demands on Water Resources, Report to Congress on the Interdependency of Water and Energy, DOE, December 2006.

[22] Gutowski, W.J., G.C. Hegerl, G.J. Holland, T.R. Knutson, L.O. Mearns, R.J. Stouffer, P.J. Webster, M.F. Wehner, and F.W. Zwiers, 2008: Causes of observed changes in extremes and projections of future changes. In: Weather and Climate Extremes in a Changing Climate: Regions of Focus: North America, Hawaii, Caribbean, and U.S. Pacific Islands [Karl, T.R., G.A. Meehl, C.D. Miller, S.J. Hassol, A.M. Waple, and W.L. Murray (eds.)]. Synthesis and Assessment Product 3.3. U.S. Climate Change Science Program, Washington, DC, pp. 81-116.

[23] Bull, S. R., D. E. Bilello, J. Ekmann, M. J. Sale, and D. K. Schmalzer, 2007: Effects of climate change on energy production and distribution in the United States in Effects of Climate Change on Energy Production and Use in the United States. A Report by the U.S. Climate Change Science Program and the Subcommittee on Global Change Research. Washington, DC.

[24] Ibid.

[25] Bates, B.C., Z.W. Kundzewicz, S. Wu, and J.P. Palutikof (eds.), 2008: Climate Change and Water. Technical paper of the Intergovernmental Panel on Climate Change (IPCC). IPCC Secretariat, Geneva, Switzerland, available at http://www.ipcc.ch/pdf/technical-papers/climate-change-water-en.pdf, accessed June 29, 2011.

[26] The National Inventory of Dams (NID) lists more than 82,000 dams, about 65 percent of which are privately owned. The total number of dams in the Nation, including those not on the NID, is estimated at 100,000 according to the 2010 Dams Sector Annual Report.

[27] The Columbia River basin experienced low water years during this period. See Section 3 for further discussion.

[28] Note that the Bureau of Reclamation, the owner and operator of the Grand Coulee Dam, lists that the total generating capacity of the dam as 6,809 MW. See http://www.usbr.gov/pn/grandcoulee/index.html.

[29] Wind Power: Capacity Factor and Intermittency, Council of European Municipalities and Regions http://www.ceere.org/rerl/about wind/RERL Fact Sheet 2a Capacity Factor.pdf, accessed March 17, 2011; annual average capacity factor derived from EIA data.

[30] Derived from EIA-906, EIA-920, EIA-923, and EIA-860 databases, annual electric power generation, http://www.eia.doe.gov/cneaf/electricity/page/eia906 920.html, and summer capacity, http://www.eia.doe.gov/cneaf/electricity/page/eia860.html, both accessed November 28, 2010.

[31] Martin-Amouroux, Jean-Marie, the economics of hydroelectricity, Hydro 21, Grenoble European Center on Hydropower, June 2004, http://www.hydro21.org/div media/pdf/pdf economie en.pdf, accessed November 12, 2010.

[32] The term Columbia River System used in this report encompasses the tributaries of the Columbia River and the Snake River, as seen in figure 6.

[33] ―The Columbia River System: Inside Story," Second Edition, Bonneville Power Administration (BPA), April 2001, http://www.bpa.gov/power/pg/columbia river inside story.pdf, accessed December 17, 2010.

[34] Federal Columbia River Power System (FCRPS), BPA, http://www.bpa.gov/power/pgf/hydrpnw.shtml, accessed February 17, 2011.

[35] On average, about 25 percent of the Columbia River flow comes from Canada. See ―the Columbia River System: Inside Story."

[36] Ibid., and 2009 BPA Facts, BPA, http://www.bpa.gov/corporate/about BPA/Facts/FactDocs/2009 BPA Facts. pdf, accessed January 6, 2011.

[37] ―House panel to probe impacts of regulations on hydropower,‖ Environment & Energy Daily, March 16, 2011. 381997 Pacific Northwest Coordination Agreement (PNCA), June 18, 1997, http://www.nwd-wc.usace.army .mil/PB/oper planning/97PNCA Conformed.pdf, accessed January 25, 2011.

[39] Bates, B.C., Z.W. Kundzewicz, S. Wu, and J.P. Palutikof (eds.), 2008: Climate Change and Water. Technical paper of the Intergovernmental Panel on Climate Change. IPCC Secretariat, Geneva, Switzerland, available at http://www.ipcc.ch/pdf/technical-papers/climate-change-water-en.pdf, accessed June 29, 2011.

[40] Stewart, I.T., D.R. Cayan, and M.D. Dettinger, 2004: Changes in snowmelt runoff timing in western North America under a ‗business as usual' climate change scenario. Climatic Change, http://meteora.ucsd.edu/cap/stewart clch.pdf, accessed June 29, 2011.

[41] Ingram, H., D. Feldman, N. Mantua, K.L. Jacobs, D. Fort, N. Beller-Simms, and A. M. Waple, 2008: The Changing Context. In: Decision-Support Experiments and Evaluations using Seasonal-to-Interannual Forecasts and Observational Data: A Focus on Water Resources. A Report by the U.S. Climate Change Science Program and the Subcommittee on Global Change Research [Nancy Beller-Simms, Helen Ingram, David Feldman, Nathan Mantua, Katharine L. Jacobs, and Anne M. Waple (eds.)]. NOAA's National Climatic Data Center, Asheville, NC, pp. 7-28.

[42] Kevin McCullen, ―Low snowpack to cost BPA projected $233M," Tri-city Herald, May 4, 2010, http://www.tri-cityherald.com/2010/05/04/1001094/low-snowpack-to-cost-bpa-projected.html, accessed November 20, 2010.

[43] 2010 Annual Report, BPA, http://www.bpa.gov/corporate/finance/a report/10/AR2010.pdf, accessed January 4, 2011.

[44] Jim Mann, the Western News, November 16, 2010, http://www.thewesternnews.com/news/article 3435b546-f1ea-11df-b434-001cc4c002e0.html?mode=print, accessed December 30, 2010.

[45] ―Columbia River high-water operations, [June 1-14, 2010],‖ BPA, September 2010, http://www.bpa.gov/corporate/pubs/final-report-columbia-river-high-water-operations.pdf, accessed March 15, 2011.

[46] Excessive spill is restricted because it can produce very high concentrations of total dissolved gas in the water that can cause gas bubble trauma in fish. See Statement on Environmental Redispatch and Negative Pricing, BPA, December 3, 2010, http://www.bpa.gov/corporate/AgencyTopics/ColumbiaRiver HighWaterMgmnt/Environmental%20Redispatch%20s tatement.pdf, accessed March 15, 2011.

[47] Lack-of-market spill means water that could have been used to generate power and reduce excess spill had a market for power and/or transmission to reach market been available. See ―Columbia River high-water operations, [June 1-14, 2010],"BPA report.

[48] Ibid.

[49] ―Columbia River high-water operations, [June 1-14, 2010]," BPA report.

[50] Columbia River high water management, BPA, http://www.bpa.gov/corporate/AgencyTopics/ColumbiaRiverHighWaterMgmnt/, accessed March 15, 2011.

[51] BPA's Interim Environmental Redispatch and Negative Pricing Policy, Administrator's Final Record of Decision, May 2011, BPA, http://www.bpa.gov/corporate/pubs/RODS/2011/ERandNegativePricing FinalROD web.pdf, accessed July 7, 2011.

[52] Colorado River Law and Policy, Western Water Assessment, http://wwa.colorado.edu/colorado river/law.html, accessed December 12, 2010.

[53] Pontius, Dale, ―Colorado River Basin Study: Final Report," The Western Water Policy Review Advisory Commission, August 1997, http://wwa.colorado.edu/colorado river/docs/pontius%20colorado.pdf, accessed December 12, 2010.

[54] The Law of the River, Bureau of Reclamation, http://www.usbr.gov/lc/region/g1000/lawofrvr.html, accessed December 16, 2010.

[55] Colorado River Compacts, The Colorado River Water Conservation District, http://www.crwcd.org/media/uploads/20080416 CO River Compact.pdf, accessed December 16, 2010.

[56] Pontius, Dale, ―Colorado River Basin Study: Final Report," 1997.

[57] U.S. Census Bureau, http://quickfacts.census.gov/qfd/maps/thematic/PL0120000.html, accessed December 16, 2010.

[58] ―Colorado River Basin Water Management: Evaluating and Adjusting to Hydroclimatic Variability," Committee on the Scientific Bases of Colorado River Basin Water Management, National Research Council, 2007, http://www.nap.edu/catalog/11857.html, accessed January 25, 2011.

[59] Ibid.; These include a statewide water supply initiative in Colorado in 2004, Arizona Drought Preparedness Plan of 2004, as well as the incorporation of climate change into water management in California in 2006.

[60] Pontius, Dale, ―Colorado River Basin Study: Final Report,‖ 1997.

[61] ―Colorado River Interim Guidelines for Lower Basin Shortages and Coordinated Operations for Lakes Powell and Mead," Bureau of Reclamation, http://www.usbr.gov/lc/region/programs/strategies/about.html, accessed December 17, 2010.

[62] Scoping Summary Report for Colorado River Interim Guidelines, Appendix V Summary of Issues Raised in Comments Grouped by Resource/Issue Area, Bureau of Reclamation, March 2006, http://www.usbr.gov/lc/region/programs/strategies/scopingreport/Appendices/AppV.pdf, accessed October 13, 2011.

[63] The Record of Decision, Colorado River Interim Guidelines for Lower Basin Shortages and Coordinated Operations for Lakes Powell and Mead, Bureau of Reclamation, http://www.usbr.gov/lc/region/programs/strategies/RecordofDecision.pdf, and ―Secretary Kempthorne Signs Historic Decision For New Colorado River Management Strategies," U.S. Department of Interior, December 13, 2007, http://www.doi.gov/archive/news/07 News Releases/071213.html, accessed January 27, 2011.

[64] Eric Wolff, ―ENERGY: Hoover Dam could stop generating electricity as soon as 2013, official fear," North County Times, September 11, 2010, http://www.nctimes.com/business/article b7e44e9e-087d-53b2-9c49-7ea32262c9a9.html, accessed November 20, 2010.

[65] Janet Zimmerman, ―REGION: Colorado River drought threatens power production," The Press-Enterprise, October 6, 2010, http://www.pe.com/localnews/stories/PE News Local D mead06.2b8c2c6.html, accessed January 4 2010.

[66] Upper Colorado Region, Bureau of Reclamation, http://www.usbr.gov/uc/water/crsp/cs/gcd.html, accessed June 29, 2011.

[67] Ibid.

[68] ―Colorado River Basin Water Supply and Demand Study Plan of Study," Bureau of Reclamation, http://www.usbr.gov/lc/region/programs/crbstudy.html, accessed May 5, 2011.

[69] James Cartledge, ―Hoover Dam turbines set for upgrade to cope with drought,‖ Brighterenergy.org, April 19, 2010, http://www.brighterenergy.org/9098/news/marine-hydro/hoover-dam-turbines-set-for-upgrade-to-cope-withdrought/, accessed November 20, 2010.

[70] Annual Operating Plans, Upper Colorado Region, Bureau of Reclamation, http://www.usbr.gov/uc/water/rsvrs/ops/aop/index.html, accessed December 16, 2010.

[71] Record of Decision, Operation of Glen Canyon Dam, Final Environmental Impact Statement, Bureau of Reclamation, October 8, 1996, http://www.usbr.gov/uc/rm/amp/pdfs/sp appndxG ROD.pdf, accessed October 13, 2011.

[72] 2007 TVA Strategic Plan, Tennessee Valley Authority (TVA), http://www.tva.com/stratplan/tva strategi c pl an.pdf, accessed December 21, 2010.

[73] Tennessee River and Reservoir System Update, TVA, December 6, 2010, http://www.tva.gov/email /eRiver/2010/december.html, accessed December 21, 2010.

[74] Ibid.

[75] Dave Flessner, ―Tennessee Valley Authority boosts rates 20 percent," Chattanooga Times Free Press, August 21, 2008, http://www.timesfreepress.com/news/2008/aug/21/tennessee-valley-authority accessed November 20, 2010.

[76] Ibid.

[77] Drought Management Plan, Tennessee Department of Environment and Conservation, February 2010, http://www.tn.gov/environment/dws/pdf/droughtmgtplan.pdf, accessed January 26, 2011.

[78] Tennessee River and Reservoir System Update, TVA, December 6, 2010, http://www.tva.gov/email/eRiver/2010/december.html, accessed December 21, 2010.

[79] Water Use Conflicts in the West: Implications of Reforming the Bureau of Reclamation's Water Supply Policies, Congressional Budget Office, August 1997.

[80] ―What is Drought?," NOAA, http://www.wrh.noaa.gov/fgz/science/drought.php?wfo=fgz, accessed March 1, 2011.

[81] CIPAC is an operational mechanism that allows the discussion of sensitive information between public and private sectors by providing a shield from the Federal Advisory Committee Act requirements.

[82] Critical Infrastructure Partnership Advisory Council, DHS, http://www.dhs.gov/files/committees/editorial 0843.shtm, accessed January 28, 2011.

[83] Emphasis added.

[84] Scott Simpson, "BC Hydro faces severe drought on Peace River," The Vancouver Herald, July 15, 2010, http://www.vancouversun.com/news/Hydro+faces+severe+drought+Peace+River/3274186/story.html, accessed November 20, 2010.

[85] Justin Rubner, "Drought hits Hydropower," Atlanta Business Chronicle, November 16, 2007, http://atlanta.bizjournals.com/atlanta/stories/2007/11/19/story2.html, accessed November 20, 2010.

[86] Manitoba Hydro 2002-2004 Drought Risk Management Review, January 15, 2005, http://www.hydro.mb.ca/regulatory affairs/electric/gra 08 09/information requests/Appendix 43 Report on 2002-2004 Drought.pdf, accessed April 4, 2011.

[87] Frequently Asked Questions: Power Marketing, Western Area Power Administration, http://www. wapa.gov/about/faqpm.htm, accessed November 20, 2010.

[88] Electricity Impacts of the Drought, State of Washington, Department of Community, Trade, and Economic Development, May 9, 2005, http://www.commerce.wa.gov/ cted/documents/ID 2015 Publications.pdf, accessed November 20, 2010.

[89] Closed-loop cooling systems that use cooling ponds withdraw cool water from the bottom of the ponds to dissipate the heat and cool the water. Nonetheless, some evaporation from the cooling ponds is produced.

[90] Hightower, M., Energy and Water: Overview of Emerging Issues and Challenges, DOE/EIA 2010 Energy Conference.

[91] USGS Circular 1344, United States Geological Survey, October 2009.

[92] American Waterway Operators, www.americanwaterways.com/industry stats/index.html, accessed December 16, 2010.

[93] Coal: Research and Development to Support National Energy Policy, Committee on Coal Research, Technology, and Resource Assessment to Inform Energy Policy, National Research Council, 2007.

[94] Lynn Muench, American Waterways Operators, quotes in New York Times, August 15, 2005, http://www.nytimes.com/2005/08/15/national/15drought.html? r=1, accessed December 16, 2010.

[95] Ibid.

[96] Impact of Drought on U.S. Steam Electric Power Plant Cooling Water Intakes and Related Water Resource Issues, National Energy Technology Laboratory, DOE/NETL-2009/1364, April 2009, http://www.netl.doe.gov/technologies/coalpower/ewr/water/pdfs/final-drought%2520impacts.pdf, accessed November 11, 2010.

[97] USGS Circular 1344, United States Geological Survey, October 2009.

[98] Kenny, J.F., Barber, N.L., Hutson, S.S., Linsey, K.S., Lovelace, J.K., and Maupin, M.A., 2009, Estimated use of water in the United States in 2005: U.S. Geological Survey Circular 1344, 52 p.; Solley, Wayne B., Pierce, Robert R., Perlman, Howard A., 1998, Estimated use of water in the United States in 1995: U.S. Geological Survey Circular 1200.

[99] "Colorado River Interim Guidelines for Lower Basin Shortages and Coordinated Operations for Lakes Powell and Mead," Bureau of Reclamation, http://www.usbr.gov/lc/region/programs/strategies/about.html, accessed December 17, 2010.

[100] Trends in Water Use in the United States, 1950 – 2005, United States Geological Survey.

In: Dams, Drought and Energy-Water Interdependencies
Editor: Daniel L. Calzi
ISBN: 978-1-62808-942-4
© 2013 Nova Science Publishers, Inc.

Chapter 2

AN ANALYSIS OF THE EFFECTS OF DROUGHT CONDITIONS ON ELECTRIC POWER GENERATION IN THE WESTERN UNITED STATES*

Leslie Poch, Guenter Conzelmann and Tom Veselka

1. BACKGROUND

This report was funded by the U.S. Department of Energy's (DOE's) National Energy Technology Laboratory (NETL) Existing Plants Research Program. The energy-water research component of this program is focused on water use at power plants. This study complements the program's overall research effort by evaluating the availability of water at power plants under drought conditions.

During the summer and fall of 2007, a serious drought affected the southeastern United States. River flows decreased, and water levels in lakes and other impoundments dropped. In a few cases, water levels were so low that power production had to be stopped or reduced. It is likely that, in coming years, competing water demands will increase. It is also possible that climatic conditions will become warmer or at least more variable, thereby exacerbating future droughts.

This report attempts to identify the system-wide impacts on the power system that could arise from various decreases in surface water levels. Our analysis is based on a separate report by Kimmell and Veil (Kimmell and Veil 2009) that (1) evaluates the sources of cooling water used by the U.S. steam-electric power plant fleet, (2) develops a database of cooling water intake locations and depths for those plants that use surface water supplies, and (3) identifies steam-electric power plants equipped with cooling water intakes that could not function if the water levels were to drop below certain thresholds. The goal of the simulations is to quantify the impacts of such water level decreases on the generation mix, future electricity prices, and

* This is an edited, reformatted and augmented version of a National Energy Technology Laboratory publication, DOE/NETL-2009/1365 dated April 2009.

carbon dioxide (CO_2) emissions that would occur if the utility and system operators were forced to take any of those steam-electric plants out of service, or reduce their outputs, for extended periods of time.

Our analysis focuses on the Western United States. We calibrate our power system dispatch model to the year 2006 and then develop projections for future years. In this document, we report results for 2010, 2015, and 2020.

2. Methodology and Assumption

2.1. Scope and Model Resolution

We estimate future generation mix, future electricity prices, and CO_2 emissions by simulating the operations of thermal and renewable power plants in the Western Electricity Coordinating Council (WECC) system, particularly the portion of WECC that is within the United States (Figure 1). The WECC regions that we model include the Northwest Power Pool (NWPP), Rocky Mountain Power Area (RMPA), Arizona-New Mexico-Southern Nevada Power Area (AZNM), and California (CAL). We pay special attention to interdependencies among hydropower and thermal power plant operations because hydropower plants may provide up to 40% of the WECC load during years when wet hydrological conditions occur. In some water basins, such as the Colorado River System, annual hydropower generation can vary by more than a factor of five (Figure 2). Hydrology conditions affect the dispatch of the thermal system, and therefore, water use by the power sector.

Figure 1. Map of WECC system (Only United States Considered for Modeling).

Hydropower plant generation is determined on an hourly time step. In the current model implementation, we simulate hydropower as an aggregate generation resource that serves both base load and peaking duties. We compile the information for the aggregation from individual plant-level data. The hourly dispatch of the aggregate power plant is based on

1. monthly generation control totals,
2. the amount of water used for base load duties, (3) estimated monthly hydropower capability, and (4) a WECC-wide hourly load profile.

For electricity demand, we construct WECC-wide hourly electricity demand profiles for 2006 to 2020 from control area load profiles in combination with forecasts from the *Annual Energy Outlook 2008* (AEO 2008) published by DOE's Energy Information Administration (EIA 2008[a]).

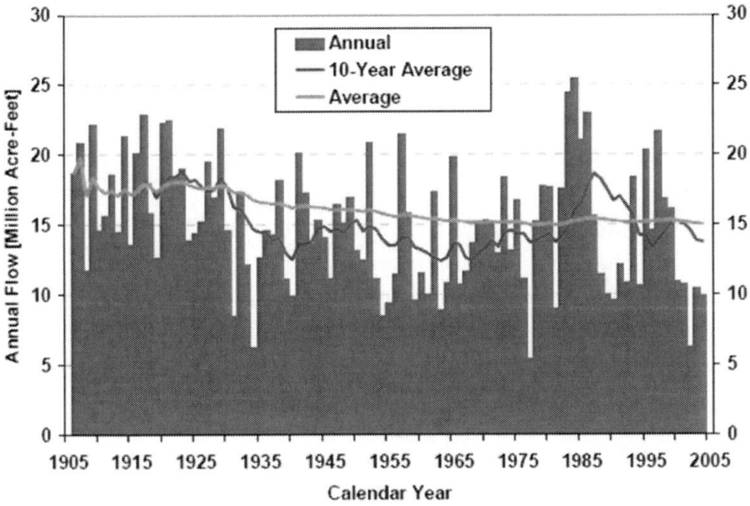

Figure 2. Example for Hydro Variability — Natural Flow for the Colorado River at Lees Ferry.

Thermal power plants are simulated at the unit level. We employ a probabilistic dispatch model to simulate thermal power plant production to meet load that is not served by hydropower plants and other renewable resources, such as wind power. We can run the thermal dispatch in two modes: by either using monthly load duration curves (LDCs) or using hourly chronological loads. In the first mode, we obtain monthly average capacity factors, generation levels, and monthly price distributions. In the second mode, we obtain hourly price distributions. In both modes, maintenance and random forced outages are accounted for at the unit level.

2.2. Analytical Process

We model the WECC–U.S. system dynamically for 2006–2020 using several modeling tools. The methodology employs the following sequence of operations:

- Collect and process data and information;
- Determine hourly renewable generation, including dispatchable and non-dispatchable aggregate hydropower and other non-dispatchable plants, such as wind;
- Determine current hourly electricity loads and forecast future load levels;

- Adjust loads for non-dispatchable renewable generation and hydropower plant generation;
- Develop baseline capacity expansion plan until 2020;
- Run a probabilistic thermal dispatch model to estimate electricity generation by thermal generation units from 2006 to 2020;
- Compute hourly prices chronologically and calculate monthly price distributions;
- Develop alternative drought scenarios;
- Run probabilistic dispatch model for the different scenarios to project hourly prices until 2020; and
- Compare and summarize results.

2.3. Data Collection and Preparation

The baseline analysis utilizes an extensive set of information. We compile the underlying data from various sources; considerable effort is spent on data validation to ensure data consistency. The following is a list of information sources used to compile the WECC-wide inventory of existing and proposed power plants, hourly load profiles, load projections, fuel price projections, and technology data.

2.3.1. Inventory of Existing and Proposed Power Plants

- Form EIA-860 (Annual Electric Generator Report) (EIA undated[a])
  - Identifies the generator location
  - Identifies the generator owner(s)
  - Provides information on summer and winter generating capability
  - Identifies the type of primary mover
  - Identifies the fuel type(s) used by the generator
- Form ETA-423 (Monthly Cost and Quality of Fuels Report) (ETA undated[a])
  - Provides information on the price of the fuel(s) used by generator
  - Provides information on the sources of the fuel(s) used by the generator
  - Provides information on the quality of the fuel(s) used by the generator (e.g., sulfur content, ash content, and higher heating value)
- Form ETA-906 (Power Plant Report) (ETA undated[a])
  - Provides information on monthly fuel consumptions by generator
  - Provides information on monthly generation levels by generator
- North American Electric Reliability Corporation (NERC) Generator Availability Data Set (GADS) (NERC 2008)
  - Provides scheduled maintenance outage rates by type of technology
  - Provides random outages by type of technology

2.3.2. Historical Load Data

- Federal Energy Regulatory Commission (FERC) Form 714 (Annual Electric Balancing Authority Area and Planning Area Report) (FERC 2009)
  - Provides information on hourly load data by control area

2.3.3. Load Projections

- WECC near-term forecast (Summary of Estimated Loads and Resources) (WECC 2007[b]) and FERC Form 714 (FERC 2009)
  - Provides information on monthly loads for two years into the future
  - Provides information on seasonal loads for 3- to 10-year forecast period
- AEO 2008 (ETA 2008[a])
  - Provides annual load projections until 2030

2.3.4. Fuel Price Projections

- AEO 2008 (ETA 2008[a]) projections
  - Provides annual fuel price escalations by fuel type until 2030

2.3.5. Expansion Candidate Technology Data

- AEO 2008 (ETA 2008[a])
  - Provides information on technical and economic performance parameters of representative power generation technologies

2.4. Treatment of Renewable Generation (Hydro and Wind)

We first estimate non-dispatchable renewable power generation. From the detailed output tables for the AEO 2008 reference case, we take the annual energy generation by renewable technology until 2020 for the three regions used by EIA to define WECC (Note: EIA divides the U.S. into 13 regions and three of those regions make up WECC, while WECC subdivides itself into four regions. EIA combines the WECC regions of AZNM and RMPA into one region, called RMPA-AZ. See Figure 10 in Section 2.7 for details). Geothermal, municipal solid waste, and wood and biomass combustion units are included in the dispatch model. For wind, we use a total of eight available wind generation patterns for the Western United States and assign them as representative wind patterns to each of the three EIA-defined regions that make up WECC to obtain hourly wind generation patterns for each WECC region. We use a scaling routine to match the AEO 2008 regional wind energy totals and sum across the regions to obtain a WECC-wide hourly wind generation trace until 2020 (Figure 3 shows Base Case WECC wind generation in 2006). This wind generation is then subtracted from the total WECC load. We complete a similar load subtraction for nondispatchable hydropower (i.e., run-of-river power plants). Section 2.6 provides more details about the load subtraction process.

To model the hourly generation pattern from dispatchable hydropower plants (plants with reservoirs or storage capabilities), we use a peak shaving approach. By using information from Form EIA-906, we estimate monthly hydropower generation patterns for individual hydropower plants. Also, data from various sources are used to separate power plant capabilities obtained from Form EIA-860 into base load and peaking duties. Total monthly hydropower generation levels and plant capabilities are then computed. Next, we simulate the hourly hydropower dispatch by using a peak shaving algorithm that minimizes the peak load that the thermal system must serve subject to monthly hydropower capacity and energy constraints, spinning reserve duties, hourly ramping constraints, and daily change limitations (Figure 4).

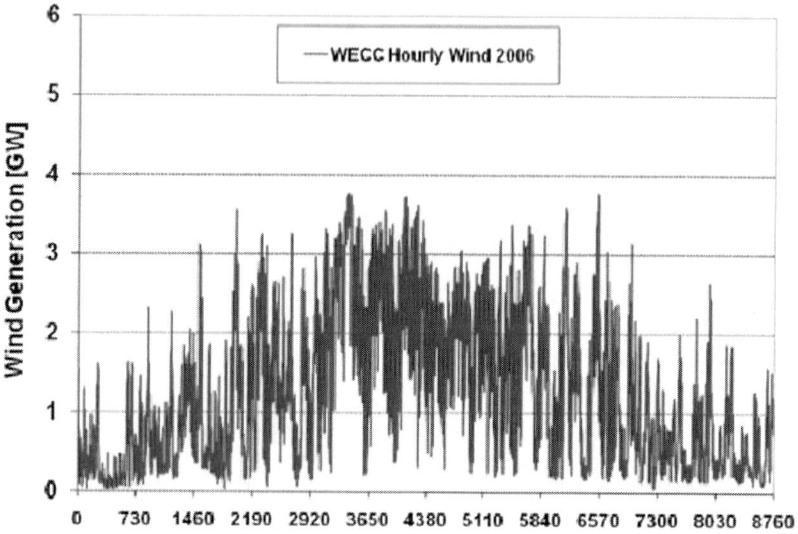

Figure 3. WECC Hourly Wind Generation 2006 .

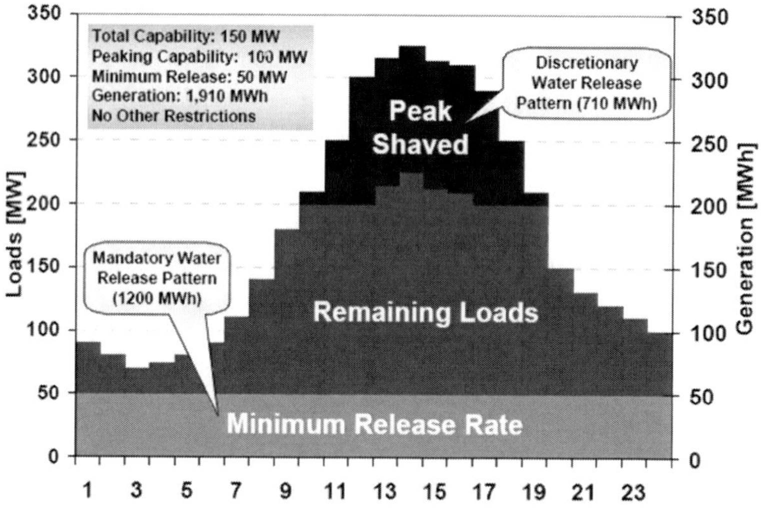

Figure 4. Hydropower Plant Operations.

2.5. Current Load and Load Forecast

Figure 5 shows the process used to develop the hourly WECC load data for the analysis period (2006–2020). First, we collect hourly historical load data for all control areas in the United States that report to WECC. We perform consistency checks on the data, making adjustments when errors are found and data are missing. Control area loads are then grouped and aggregated into the four WECC regions: NWPP, RMPA, AZNM, and CAL. The areas only cover U.S. territory. Next, we use a load-scaling algorithm to adjust aggregated hourly load profiles to exactly match the monthly peak and total load values reported for each WECC region.

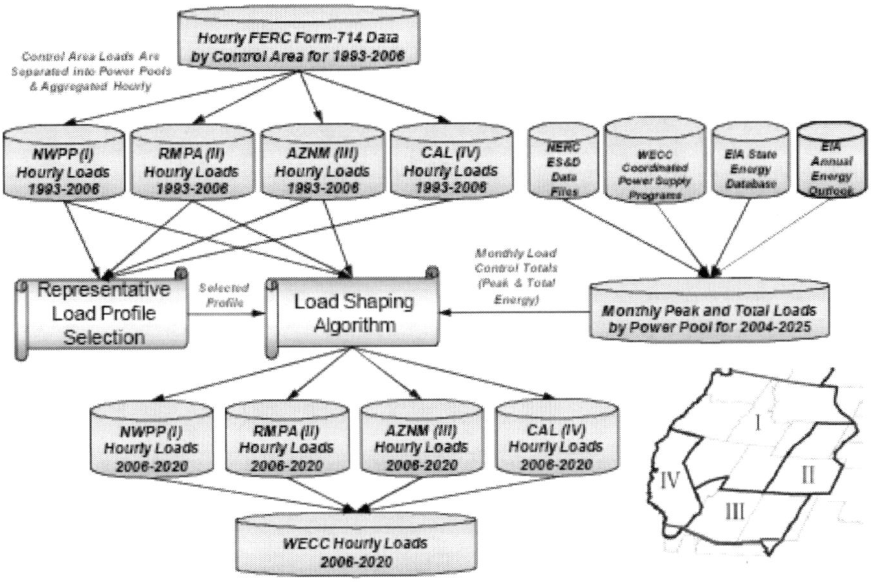

Figure 5. Processing Hourly Loads.

Figures 6 and 7 show relative monthly energy factors and monthly relative peak fractions based on FERC Form 714 for two of the major areas (RMPA and AZNM) for a selection of historical years. For each major area, we select from this data set, as the representative load profile, the data set that has the lowest sum of squared differences relative to the average profile. This representative profile is used as the basis for constructing hourly load projections for future years through 2020. The load-scaling algorithm is applied to adjust the representative hourly load profiles to match peak and total load targets that come from various statistics, including WECC's coordinated power supply programs, EIA state energy databases, EIA's AEO 2008, and the Electricity Supply and Demand (ES&D) data from the NERC.

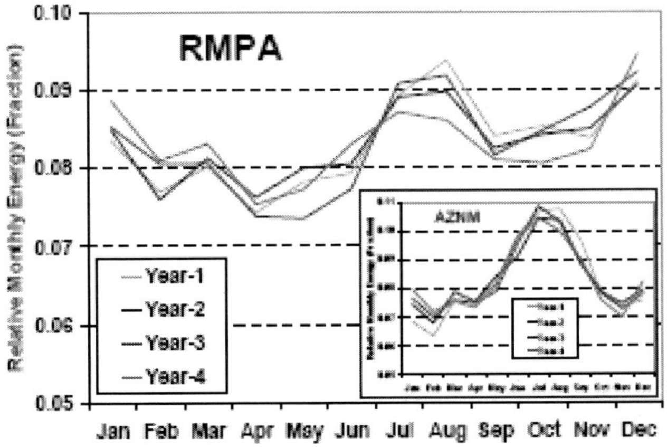

Figure 6. RMPA and AZNM Monthly Energy Factors.

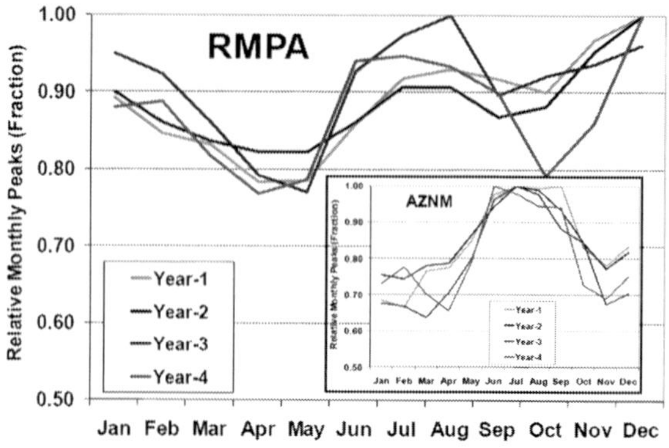

Figure 7. RMPA and AZNM Monthly Peak Factors.

2.6. Load Adjustments

As discussed in Section 2.4, the original hourly total-WECC load data series are adjusted in two ways:

1. For non-dispatchable resources (e.g., wind, run-of-river hydro) by using load subtraction.
2. For dispatchable hydropower using the peak shaving algorithm.

The remaining adjusted hourly loads are used to construct monthly LDCs that are served by the thermal system and are input into the probabilistic thermal dispatch model for the simulations. Figure 8 shows a 1-week example of how the load adjustments affect the total load served by the thermal system. Figure 9 shows the monthly load duration curves.

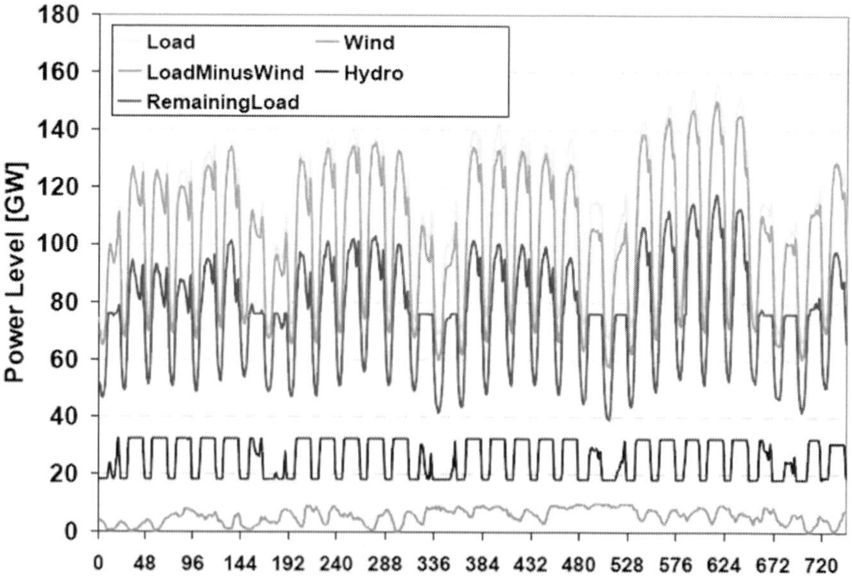

Figure 8. Example for Load Adjustment (WECC May 2020).

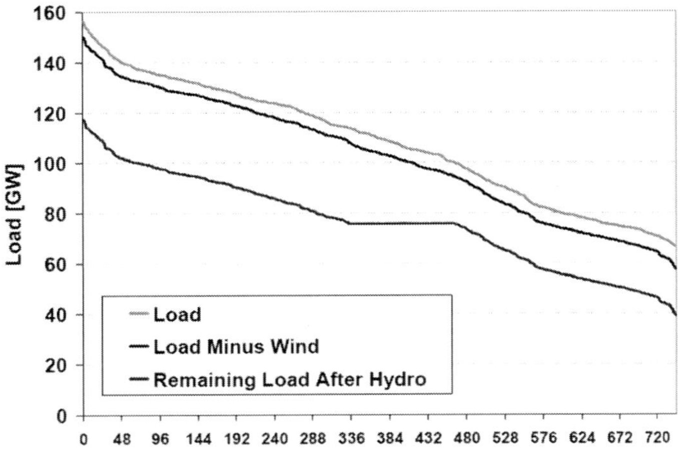

Figure 9. Example for Load Duration Curves (WECC May 2020).

2.7. Capacity Expansion Modeling

We develop the baseline capacity expansion scenario for the WECC system until 2020 by using the EIA's AEO 2008 as a starting point. EIA derives these projections by using the National Energy Modeling System (NEMS) Electricity Market Module (EMM). On the basis of the fuel prices and electricity demands provided by other modules of the NEMS, the EMM determines the most economical way to supply electricity, subject to environmental and operational constraints. A detailed description of the EMM is available in *Electricity Market Module of the National Energy Modeling System 2006* (EIA 2006).

The AEO 2008 contains projections of new capacity additions by technology for a total of 13 regions, shown in Figure 10.

Three of these regions represent a geographic area in the United States that is served by WECC:

- Region 11: Northwest Power Pool
- Region 12: Rocky Mountain Power Area, Arizona, New Mexico, and Southern Nevada
- Region 13: California

It should be noted that WECC defines four general load areas or regions within its service territory:

- Northwest Power Pool Area
- Rocky Mountain Power Area
- Arizona – New Mexico – Southern Nevada Power Area
- California – Mexico Power Area

To maintain consistency with the AEO 2008, our analysis uses a representation of the WECC system with three regions for the development of the revised capacity expansion plan.

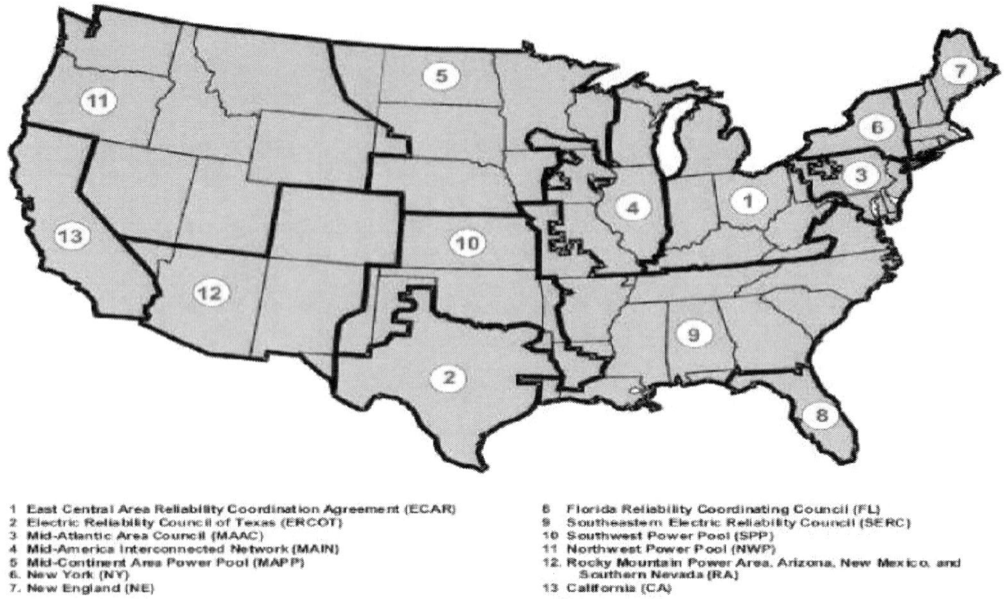

Figure 10. Annual Energy Outlook 2008 Electricity Market Model Supply Regions (Source: EIA 2008[b]).

The EMM analysis for the AEO 2008 considers a number of different candidate generating technologies. As shown in Table 1, they include both conventional and renewable technologies.

The EMM analysis also allows for changing and improving technical and economic parameters over time (i.e., learning parameters).

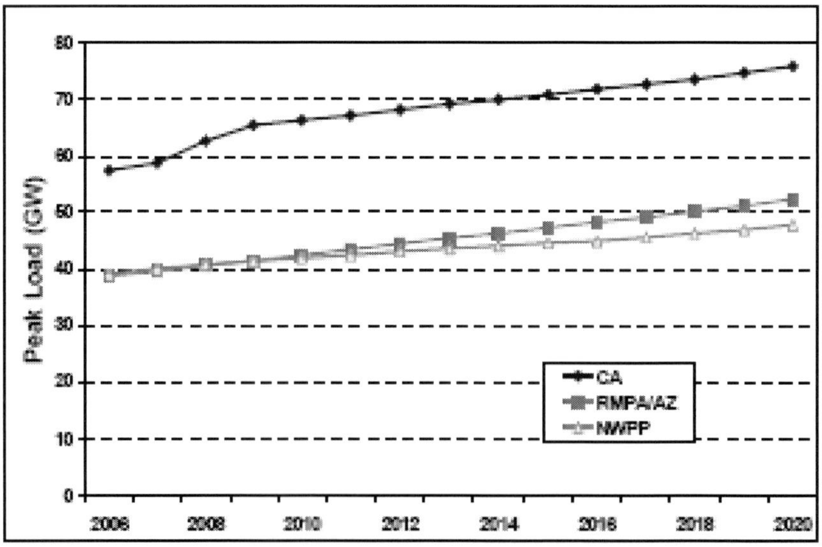

Figure 11. Annual Peak Load Forecasts until 2020.

Table 1. Generating Technologies Represented in the Electricity Market Module (Source: EIA 2008[b])

Capacity Type
Existing coal steam plants
High sulfur pulverized coal with wet flue gas desulfurization
Advance coal - integrated coal gasification combine cycle
Advanced coal with carbon sequestration
Oil/gas steam - oil/gas steam turbine
Combined cycle - conventional gas/oil combined cycle combustion turbine
Advanced combined cycle - advanced gas/oil combined cycle combustion turbine
Advanced combined cycle with carbon sequestration
Combustion turbine - conventional combustion turbine
Advanced combustion turbine - steam injected gas turbine
Molten carbonate fuel cell
Conventional nuclear
Advanced nuclear - advanced light water reactor
Generic distributed generation - baseload
Generic distributed generation - peak
Conventional hydropower - hydraulic turbine
Pumped storage - hydraulic turbine reversible
Geothermal
Municipal solid waste
Biomass - integrated gasification combined cycle
Solar thermal - central receiver
Solar photovoltaic - single axis flat plate
Wind

On the basis of the revised demand forecast for the WECC regions, we use a planning reserve margin of 15% as a driver for new capacity additions until 2020. As stated in the *WECC 2007 Power Supply Assessment* (WECC 2007[a]), the capacity needs are determined at the level of WECC regions, and each region needs to maintain a minimum planning reserve margin of 15%.

Because the reserve margin requirement is normally based on the net available capacity, while the AEO 2008 lists only installed capacities, we have increased the requirement for the NWPP region to 25% of installed capacity to account for the large amount of hydro capacity in this region.

The reserve margin requirements for the other two regions, RMPA/AZ and CAL, remain at 15%. We then perform an expansion analysis for each region individually. Therefore, the total capacity additions for the WECC system are obtained as the sum of new capacity additions in each of the regions. The overall resulting reserve margin, based on the installed capacity, for the WECC system as a whole, amounts to about 25% in 2012, gradually decreasing to about 21.4% in 2020.

The technology mix of new generating capacity until 2020 is based on the AEO 2008 projections for each WECC region. Compared with the AEO 2008 expansion plan, the 25% planning reserve margin requirement does not produce any changes in the capacity needs for the NWPP region, while the 15% reserve margin requirement requires some new generating capacity to be added to the system in addition to that already projected by the AEO 2008. For the RMPA/AZ region, this results in only slightly increased capacity needs beginning in 2019 and amounting to a cumulative total of 1,160 MW by 2020. For the CAL region, the 15% reserve margin requirement results in additional capacity needs beginning in 2012 and amounting to a cumulative total of 9,850 MW by 2020. Again, it is assumed that the technology mix for this additional capacity corresponds to that of the AEO 2008.

2.8. Thermal Dispatch Modeling

The first step in the dispatch modeling is to create a validated unit inventory for the entire WECC system. As shown in Figure 12, we use Form EIA860 as a starting point,

Form EIA-423 to add fuel data to the inventory, Form EIA-906 to obtain estimates for heat rates, the GADS database on outage information, and the AEO 2008 tables for variable operation and maintenance (O&M) costs.

With the complete unit inventory, we run a unit-level hourly thermal probabilistic dispatch model that accounts for forced outages, as well as scheduled maintenance. We estimate future maintenance schedules by using a routine that maximizes the minimum reserve margin.

Figure 13 shows sample results for the maintenance scheduler in combination with a forced outage scenario. The dispatch model utilizes a convolution process in which the loads that a unit serves include (1) the original LDC and (2) loads that could not be served by units loaded before it because of forced outages.

From the dispatch routine, we obtain unit-level generation levels, chronological prices, price distributions, and CO_2 emissions and summarize them for each simulation month. Hydropower plants in this analysis are modeled as an aggregate generation resource that serves base load and peaking duties.

The hourly dispatch of the aggregate power plant is based on monthly generation control totals, the amount of water used for base load duties, estimated monthly hydropower capability, and a WECC-wide hourly load profile.

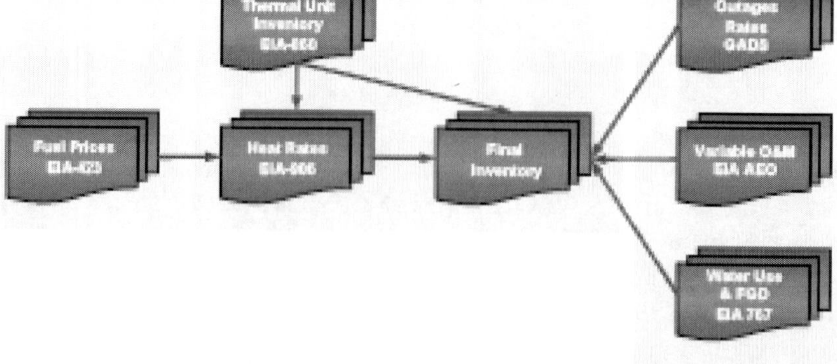

Figure 12. Creating a Thermal Unit Inventory.

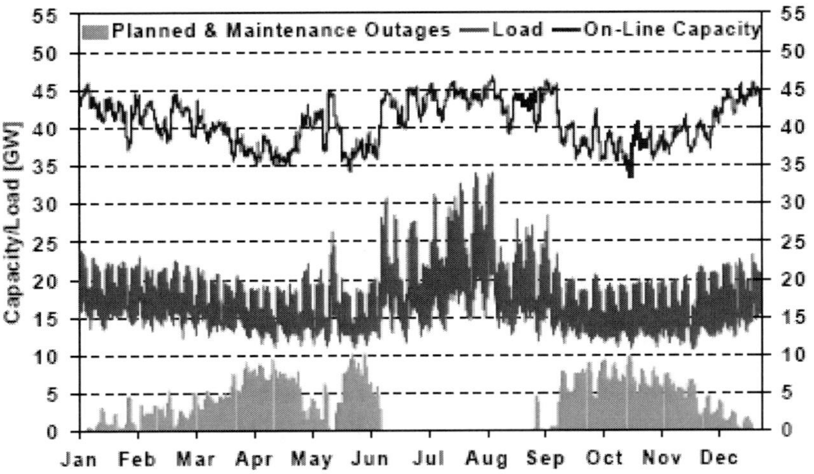

Figure 13. Example for Results of Maintenance Scheduling Routine.

3. Model Results

3.1. Base Year Model Calibration

We use information for 2006 to calibrate the model to actual observed WECC market data. Table 2 provides a comparison of model results with actual annual generation and fuel consumption data by fuel type.

Table 2. 2006 Model Calibration for Generation Mix

Technology	Model Generation Mix (%)	Actual Generation Mix (%)
Coal	31.5	31.2
Gas	26.2	25.6
Nuclear	10.4	9.4
Hydro	28.1	28.8
Wind	1.6	1.4
Others	2.2	3.6
Total	100	100

Note: Actual generation mix is calculated based on AEO 2008.

In addition to generation and fuel consumption levels, we test and calibrate the model with regard to historical prices, collecting prices from the following hubs in WECC for several historical years: Palo Verde, Pinnacle Peak, 4Corners, Mona, Mead, COB, NP15, SP15, MidColumbia, NOB, and WestWing. Prices are available in off-peak and on-peak blocks. We adjust the data set to account for the fact that off-peak prices are for 8-hour blocks, on-peak prices are for 16-hour blocks, and prices on Sundays are for 24-hour blocks. WECC system holidays are considered off-peak. From the hub prices, we calculate an average WECC system price that we compare with our modeled unconstrained system marginal price. Figures 14 through 17 show the results of the calibration process. The red bars show the price probability distributions from our model runs for each month in 2006. The blue lines show the monthly probability distribution of the estimated average WECC system price derived from daily hub prices for 2006.

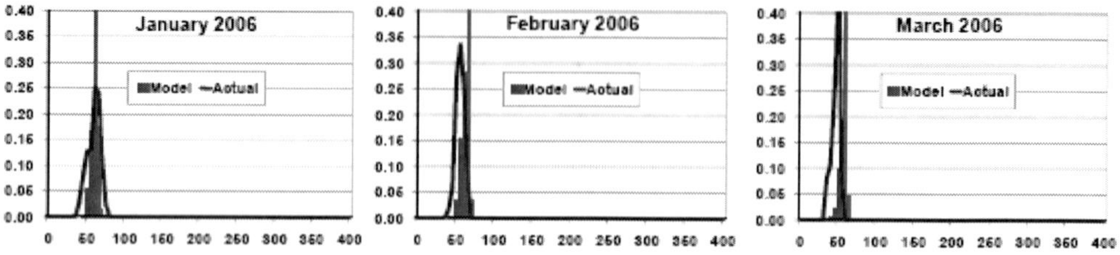

Figure 14. Model Calibration — Price Probability Distributions for January through March 2006.

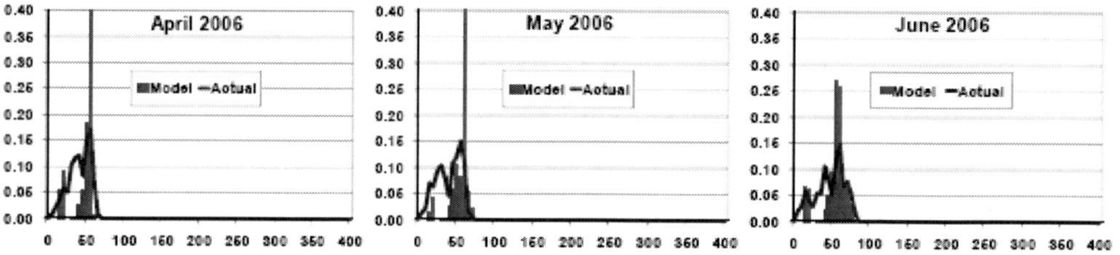

Figure 15. Model Calibration — Price Probability Distributions for April to June 2006.

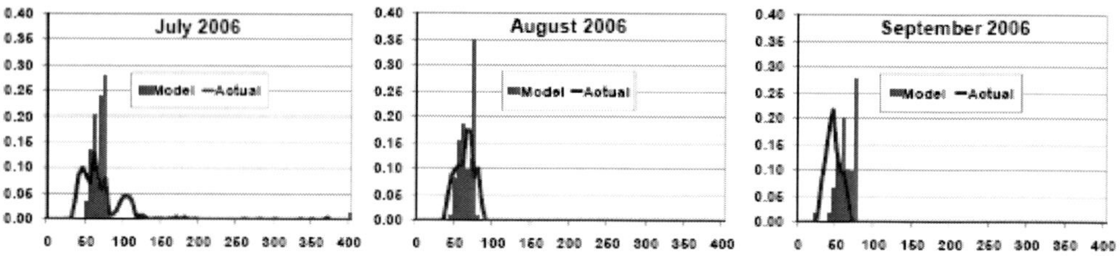

Figure 16. Model Calibration — Price Probability Distributions for July to September 2006.

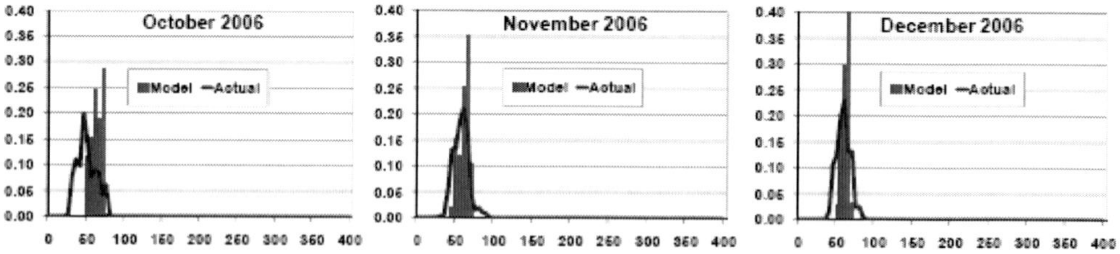

Figure 17. Model Calibration — Price Probability Distributions for October to December 2006.

3.2. Baseline Results

3.2.1. Load Projection

We project that electricity demand in the WECC–U.S. system will increase from about 700 TWh in 2006 to over 930 TWh in 2020 with a corresponding growth in peak load —

from over 135 GW to almost 170 GW over the same period. With this growth in load, the expected retirement of approximately 7.8 GW of existing generating units, and the need to maintain an adequate planning reserve margin, we foresee a need to bring online new capacity on the order of 50 GW by 2020. Figure 18 shows the capacity-load balance for the WECC system, illustrating the development of existing and new generating capacity versus the peak load until 2020.

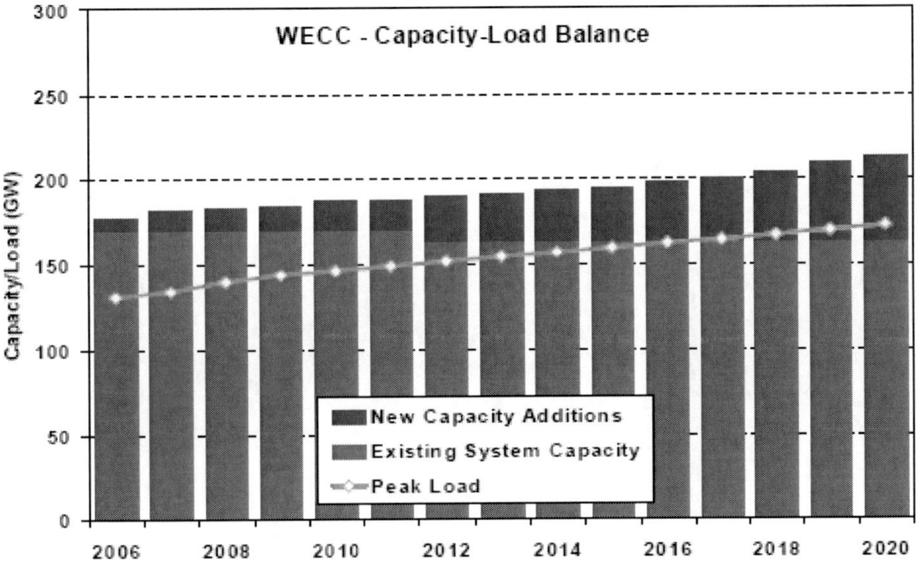

Figure 18. Baseline Projected Load, Existing System, and New Capacity Additions until 2020.

3.2.2. Capacity and Generation Projection

Figure 19 illustrates the development of generating capacity by technology type over the projection period. Total installed capacity grows from 177 GW in 2006 to 212 GW in 2020. Fuel oil capacity drops from 20 to 14.6 GW. Existing nuclear units will be allowed to retire according to schedule with no new nuclear capacity assumed to come online during the study period in the WECC system. Major growth is projected for coal and renewables, with increases from 32 to 59 GW and 59 to 66 GW, respectively, with an additional 350 MW of small distributed generation capacity.

Figure 20 shows the technology mix of the new capacity additions. By 2020, a total of 50 GW of new capacity is projected to come on line. Coal takes the largest share with 27 GW (55% of total additions), followed by 14 GW of gas-fired units (29%), and 8 GW of renewables and small distributed generators (16%). We also assume that new coal plants will be equipped with a cooling system that will be much less vulnerable to drought conditions, such as dry cooling (which requires little or no water).

Figure 21 provides a breakdown of renewable capacity for the WECC system until 2020. Conventional hydro capacity essentially stays flat at around 52 GW. Geothermal and wind increase from 2.4 to 3.1 GW and from 5.1 to 8.1 GW, respectively. Most of the renewable capacity additions come from hydropower (2 GW), wind (3 GW), and geothermal (0.7 GW), with the balance coming from smaller amounts of solar thermal and solar photovoltaic (PV), municipal solid waste, and wood/biomass.

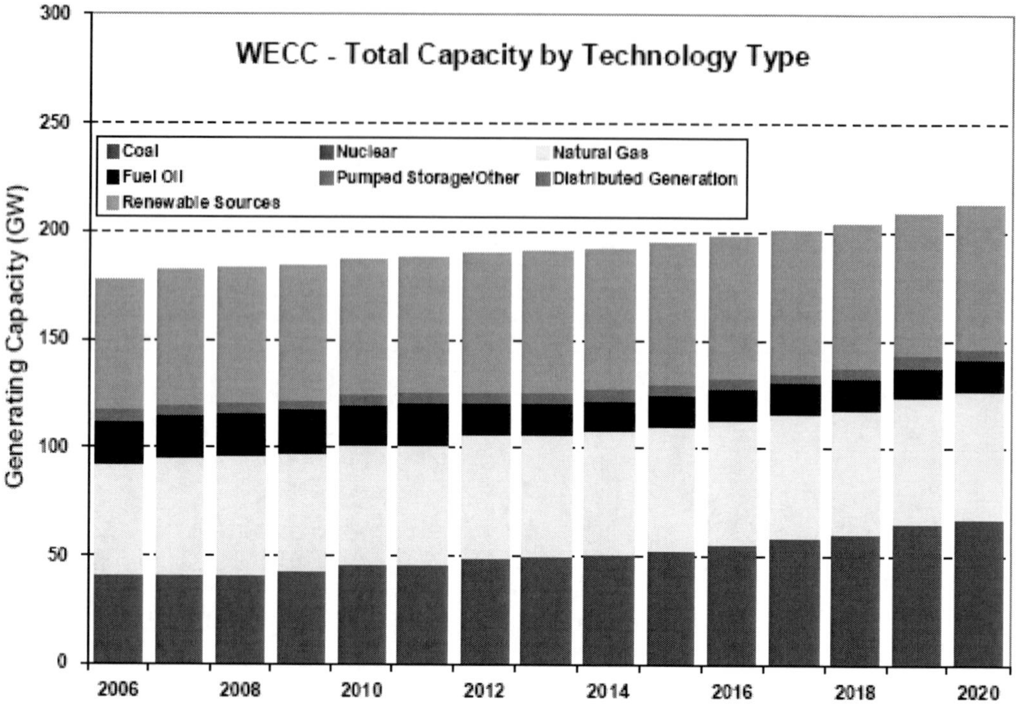

Figure 19. Baseline Projected Total Installed Capacity until 2020.

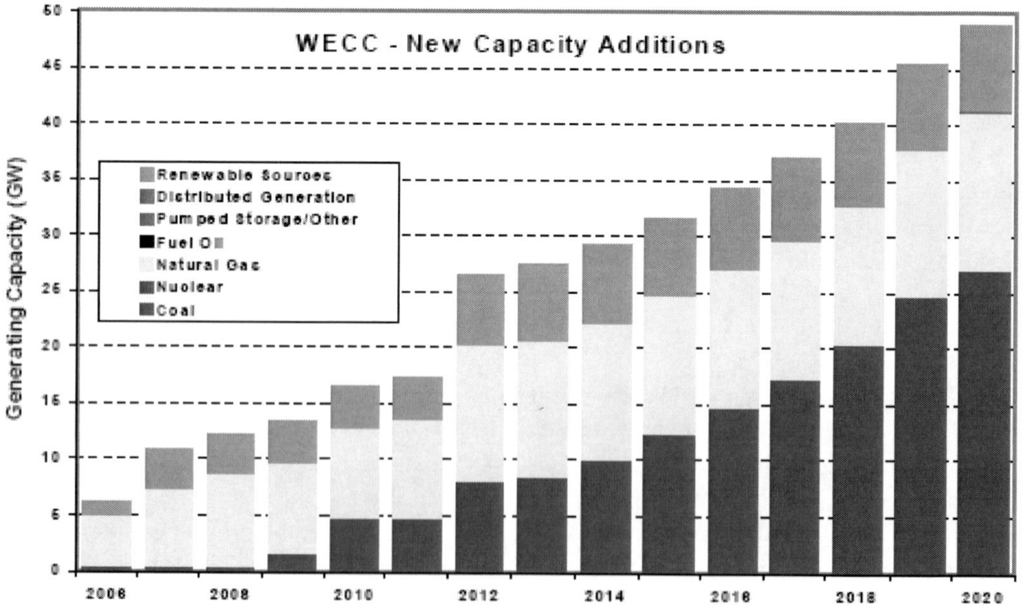

Figure 20. Baseline Projected Capacity Additions until 2020.

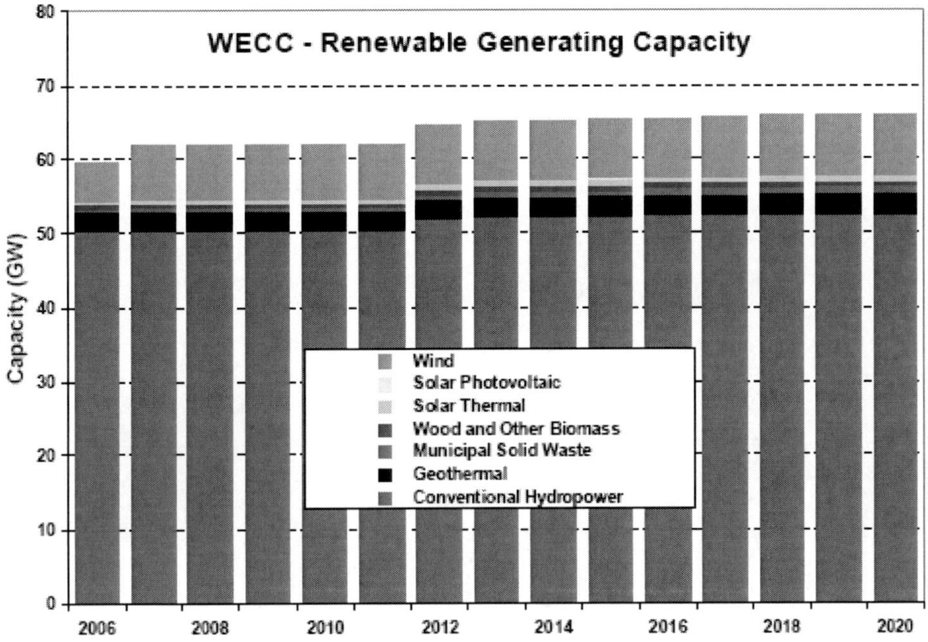

Figure 21. Baseline Projected Total Installed Renewable Capacity until 2020.

3.2.3. CO_2 Emissions Projection

Carbon dioxide emissions result from th ombustion of fuelse ccontaining carbon. In this study, the carbon-based fuels are coal, natural gas, fuel oil, and biomass. Because CO_2 emissions from biomass are highly dependent upon its composition, and because biomass makes up only about 1% of the generating capacity in the western United States, emissions from biomass power plants are not addressed in this study.

For the remaining thermal plants , CO_2 emissions vary by plant and depend on the fuel type, the efficiency of the power plant (or heat rate [measured in Btu/kWh]), and the amount of electricity the plant produces.

Emissions of CO_2 are calculated by using an emission factor. Emission factors have been developed for all types of carbon-based fuels; they measure the amount of CO_2 released (in lb) per unit of heat (Btu) generated during combustion. Emission factors for this study were obtained from the EIA Web site and are listed in Table 3. The value for coal is the average of the emission factor for bituminous and sub-bituminous coals, which are the two types of coal used in power plants in the western United States.

Table 3. CO_2 Emission Factor by Fuel Type

Fuel Type	CO_2 Emission Factor (lb/million Btu)
Coal	209.0
Natural Gas	116.4
Heavy Fuel Oil	173.7
Light Fuel Oil	161.3

Source: EIA undated[b].

One of the results of the baseline thermal dispatch model run is the amount of electricity each plant in the unit inventory produces each month of the year. The unit database contains the efficiency or heat rate of each plant. Multiplying each plant's emission factor by the heat rate and the amount of electricity it generates in a year yields the amount of CO_2 the plant produces. Summing the CO_2 emissions from all the plants in the inventory yields the total amount of C produced by the electric power system. Table 4 lists the CO_2 emissions produced in each year of the study period.

Table 4. Amount of CO_2 Emissions for Baseline Scenario

Year	CO_2 Emissions (million short tons)
2010	408.4
2015	480.5
2020	548.1

3.3. Drought Scenario

This section discusses the major assumptions behind the drought scenario and compares the results of the thermal dispatch model runs for the baseline and drought scenarios with respect to generation mix, electricity prices, and CO_2 emissions.

3.3.1. Major Scenario Assumptions

A drought would adversely impact not only thermal power plants that use fresh surface water for cooling, but also hydroelectric power plants. Hydropower production affects the load that must be served by the thermal systems, including power plants that do *not* rely on surface water. As hydropower generation is reduced as a result of drought conditions, the thermal system must operate at a higher level to compensate for lower hydropower production levels. The WECC electric grid relies very heavily on hydroelectric power. Approximately 28% of the electric power capacity is supplied by hydroelectric power plants; this percentage increases to as much as 40% in a wet hydrologic year. Therefore, to accurately simulate the effects of drought on power system operations in the WECC, we must determine the impacts of a drought on hydroelectric power generation.

In order to determine how much the amount of electricity generated by hydroelectric power plants would be reduced as the result of a severe drought, we reviewed data on hydroelectric output from 1980 and 2005. We selected 1980 as the first year of the review period because the vast majority of current WECC hydropower capacity was on line in that year, and only a very small amount of that capacity had been retired during that time period. After reviewing this hydroelectric power generation data, we selected the year with the lowest hydroelectric power production to be representative of a year in which hydropower was most affected by severe drought conditions. We assumed that the monthly amount of generation and the capacity pattern for this historic low-hydropower year would represent the operation of hydroelectric power plants in each analysis year of the drought scenario.

After determining the hydroelectric generation pattern for the drought scenario, we calculate the load pattern to be supplied by the dispatchable thermal power plants using the method described in Section 2.4 (i.e., the nondispatchable or run-of-river hydroelectric generation value is subtracted from the hourly loads remaining after wind generation is subtracted from the original WECC loads). The peak shaving algorithm is then used to model the hourly generation pattern from dispatchable hydroelectric power plants and, ultimately, to calculate the hourly loads to be supplied by thermal power plants.

The inventory of thermal power plants in the WECC system that may be adversely impacted by a drought is based on a task performed by another Argonne team and described in a separate report (Kimmell and Veil 2009). Kimmel and Veil developed a database identifying fossil and nuclear power plants equipped with cooling systems that use fresh surface water. Data included plant name, location, plant code, owner, fuel type, nameplate capacity, source of cooling water, depth of cooling water intakes, and other characteristics.

As stated in that report, drought conditions can be highly variable across the United States; they can affect large areas of the country for a long period or small areas for a short period. Because of this variability, it is highly unlikely that all of the thermal power plants using surface water for cooling would have to shut down or curtail operations in an area as large as the western United States during a drought, regardless of the depth of their water intakes. Therefore, simultaneous shutdown of all power plants in the WECC system as the result of a drought would probably be an unrealisti c scenario.

Consequently, we employ an alternative approach, using the information available on the U.S. Drought Monitor (University of Nebraska Lincoln 2009), a Web site funded by several Federal agencies and operated by the University of Nebraska Lincoln. Researchers compile and archive drought conditions on a weekly basis, from 2000 to the present, and post them on the Web site. Drought conditions are shown graphically by state and also by county within each state. For this study, we chose drought conditions for the week of January 27, 2009, to develop a plausible drought scenario and to illustrate Argonne's electric power system simulation methodology. Figure 22 shows how data are displayed on the Web site on a regional and state basis.

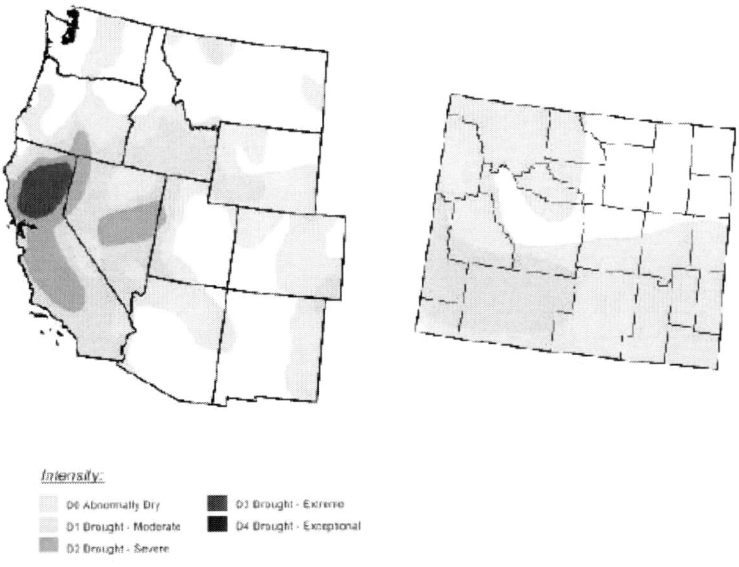

Figure 22. Sample of Data Displayed on the U.S. D roug ht Monitor Web Site— Western United States and Wyoming Drought Condition s as of Janu ary 27, 2009 (University of Nebraska Lincoln 2009).

To identify the plants that could be affected by the drought conditions during the chosen week, we compare the locations of the power plants in the WECC system with the maps on the U.S. Drought Monitor (University of Nebraska Lincoln 2009). We obtain the locations, in latitude and longitude coordinates, for each plant from the database of power plants developed by the companion Argonne study (Kimmell and Veil 2009). A geographical information system (GIS) program is used to plot the locations of the WECC power plants in the database; each location is visually compared with the state maps in the U.S. Drought Monitor. If a power plant was located in a part of the state that was designated as undergoing a moderate or more severe drought, it was chosen for shutdown or curtailment in each year of the study period.

By using this methodology, we identified a total of five plant sites in four states that would be shut down or for which operations would be curtailed. The total capacity of these plants is 3,284 MW; 2,820 MW (or 86%) of this total is supplied by coal-fired power plants. Because combined-cycle plants are very prevalent in the WECC system, their operation was handled in a special manner in this analysis. Combined-cycle plants consist of a gas turbine and a steam turbine that can be operated independently of one another, depending upon the configuration; typically, the gas turbine can operate independently of the steam turbine. The steam turbine is the only component that requires water for cooling. Therefore, in cases in which combined cycle plants were identified as possible candidates for shutdown during a drought, only the steam turbine portion of a combined-cycle unit was shut down.

3.3.2. Impacts on Generation Mix and Generation Cost

By using the technique described in Section 3.3.1, we determined the amount and generating pattern of hydroelectric power plants during a drought. Our analysis revealed that, in a severe drought year, the electrical generation from hydroelectric can drop by almost 30%. These data were input into the thermal dispatch model, and simulations were run for the two scenarios for 2010, 2015, and 2020. Table 5 and Figure 23 show model results for the amounts of electricity produced by fuel type. The amount of energy not served (ENS) is also shown. Energy not served is the amount of energy demanded by customers that the system's energy sources are unable to provide. This energy must be supplied by a source outside of the system or system operators must take steps to reduce load.

Table 5. Quantity of Electricity Generated by Fuel Type — Base and Drought Scenarios

	Base Scenario Energy (TWh)			Drought Scenario Energy (TWh)		
Fuel	2010	2015	2020	2010	2015	2020
Nuclear	74.7	74.7	74.7	74.7	74.7	74.7
Coal	257.1	314.2	417.5	236.5	293.6	401.9
Natural Gas	244.9	252.1	161.1	320.1	326.1	231.0
Fuel Oil/Other	0.90	0.90	0.78	0.91	0.93	0.86
Renewable	36.8	44.4	47.1	36.8	44.4	47.1
Hydro	186.4	185.2	185.8	131.6	131.6	131.3
ENS	0.036	0.124	0.030	0.161	0.259	0.065
Total	800.8	871.6	886.9	800.8	871.6	886.9

In the drought scenario, electricity generated from coal dropped by 20.6 TWh (about 8% compared with the baseline) in 2010, by 20.6 TWh (6.6%) in 2015, and by 15.6 TWh (3.7%) in 2020. The 30% drop in generation from hydroelectric power during a drought resulted in about 54 TWh less hydroelectric energy generated in the drought scenario. A significant increase in generation from plants using natural gas compensated for the shortfall in generation from coal and hydropower. Electricity production from natural gas rose by 75.3 TWh (30.8% compared with the baseline) in 2010, 74 TWh (29.3%) in 2015, and 70 TWh (43.5%) in 2020. Generation from other fuel sources, such as fuel oil and renewables, rose only slightly — no more than 0.1 TWh in any simulated year. Natural gas plants made up for almost the entire amount of electricity not generated by coal and hydropower.

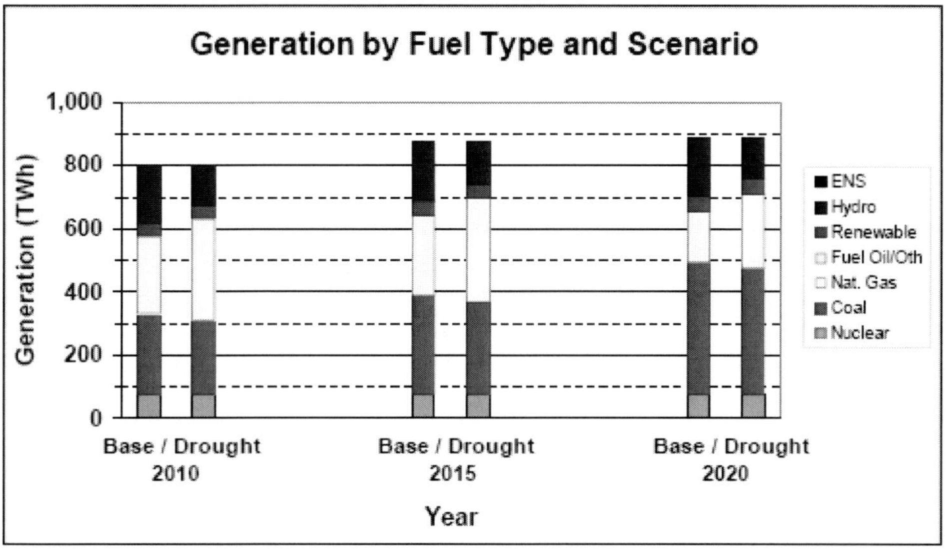

Figure 23. Electricity Generated by Fuel Type — Base and Drought Scenarios (Note: The quantity of ENS and Fuel Oil/Other cannot be seen on this plot because of its small amount compared with the amount from other sources).

The reason that natural gas plants were able to generate most of the electricity lost as a result of coal plant shutdown and the reduction in hydropower can be seen by examining their capacity factors from the base scenario model runs. The capacity factors of natural gas plants in 2010, 2015, and 2020 were 37.4%, 36.7%, and 23.1%, respectively. Because their capacity was not fully utilized, they were able to pick up the slack in the drought scenario. By 2020 though, coal's contribution starts rising, while the contribution of natural gas begins to fall. This is because coal plants with cooling technologies less vulnerable to drought are being installed in greater numbers and, by 2020, begin to displace generation from natural gas plants which, in 2010 and 2015, picked up the slack for generation from coal plants lost to drought conditions.

Nuclear power plants were unable to supply additional generation capacity in the drought scenario for two reasons: (1) no new nuclear plant came online during the study period, and (2) nuclear provides base load electricity and already generates up to its maximum potential even in the base case. There was no excess nuclear capacity to generate more electricity. In the WECC system, it is also fortunate that cooling water for nuclear power plants comes predominately from sources other than fresh surface water; otherwise, they may have been subject to the drought shutdown.

The amount of ENS increased significantly in the drought scenario, rising by more than 3.5 times in 2010 and more than doubling in 2015 and 2020. Furthermore, if ENS occurs, there is more than a 99.9% chance that it would occur in either July or August because demand for electricity in the WECC peaks during the summer months.

Because of the sharp increase in electricity produced by natural gas plants in the drought scenario, the cost to produce electricity compared with the cost in the baseline scenario increased sharp as well. T his is because operating costs of natural gas plants can be more than 3 times that of coal plants. Total electricity production costs in the baseline scenario were 17.9 billion in 2010, $17.8 billion in 2015, and $15.2 billion in 2020. Total ENS costs in the baseline scenario were $33.9 million 2010, $124 million in 2015, and $30.9 million in 2020. Figure 24 shows the differences in production costs and ENS costs between both scenarios. Production costs rose by $4.5 billion ()25.2%) in 2010, $3.9 billion (21.9%) in2015, and $33.4 million in 2020, assuming that ENS is valud at about $ 1000/MWh. This is considered a conservative value; surveys have indicated that the cost of ENS can frequently exceed $2,000/MWh (Cramton and Lien 2000).

Production costs and ENS costs decrease over time because new coal plants with cooling technologies less vulnerable to drought begin displacing generation from natural gas plants, whose generation increased in 2010 and 2015 to make up for generation lost from existing coal plants as a result of drought conditions. The new coal plants are more efficient and much less expensive to operate.

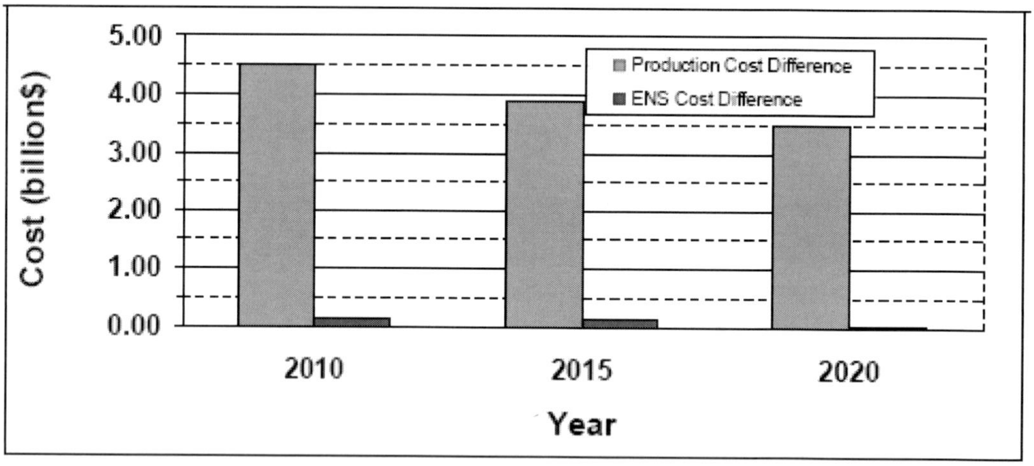

Figure 24. Production Cost and ENS Cost Differences between Base and Drought Scenarios.

3.3.3. Impacts on Electricity Prices

The thermal dispatch model generates a variety of price outputs, including monthly price distributions and hourly chronological prices with associated uncertainty ranges for user-specified percentiles.

Table 6 lists average monthly system-wide electricity prices, calculated on the basis of monthly price distributions obtained from the model.

Table 6. Average Monthly Price of Electricity — Base and Drought Scenarios

Month	Average Price of Electricity ($/MWh)						Price Difference (%)		
	Base Scenario			Drought Scenario					
	2010	2015	2020	2010	2015	2020	2010	2015	2020
Jan	61.01	54.04	51.76	65.97	58.32	56.79	8.1	7.9	9.7
Feb	60.21	53.30	50.67	67.21	59.40	54.29	11.6	11.5	7.2
Mar	55.58	49.14	46.02	60.84	53.38	50.69	9.5	8.6	10.1
Apr	54.95	48.47	43.61	61.08	53.45	50.27	11.1	10.3	15.3
May	54.69	46.88	40.57	62.23	53.06	48.29	13.8	13.2	19.0
Jun	55.35	48.71	40.04	61.80	54.96	47.48	11.7	12.8	18.6
Jul	69.14	68.07	54.17	91.67	89.16	67.24	32.6	31.0	24.1
Aug	78.48	87.87	61.75	105.70	109.75	71.27	34.7	24.9	15.4
Sep	59.97	52.85	44.95	64.05	56.73	50.17	6.8	7.3	11.6
Oct	63.20	55.75	43.04	65.47	57.86	47.24	3.6	3.8	9.8
Nov	62.97	55.36	52.13	65.89	58.18	56.36	4.6	5.1	8.1
Dec	59.44	52.70	50.89	66.72	58.71	55.30	12.2	11.4	8.7

The difference in the average price between the two scenarios is highest in the summer months (July and August), when demand in the WECC regions peaks. In 2010 and 2015, the average price for the drought scenario was 25–35% higher in those months. The difference in average prices drops considerably with time. In 2010, the average drought price in August was 35% higher than the base scenario price, but by 2020, the price was only 15% higher.

The distribution of prices is shown in Figures 25, 26, and 27 for January, a typical winter month, and August, the peak summer month. The price distribution is much larger for August compared with January for all years. In fact 5–10% of the time, prices exceed $150/MWh in August 2010 and 2015 in the drought scenario. That probability drops to about 2% in August 2020. Also, as the study progresses, the price distribution for both scenarios shifts toward lower prices.

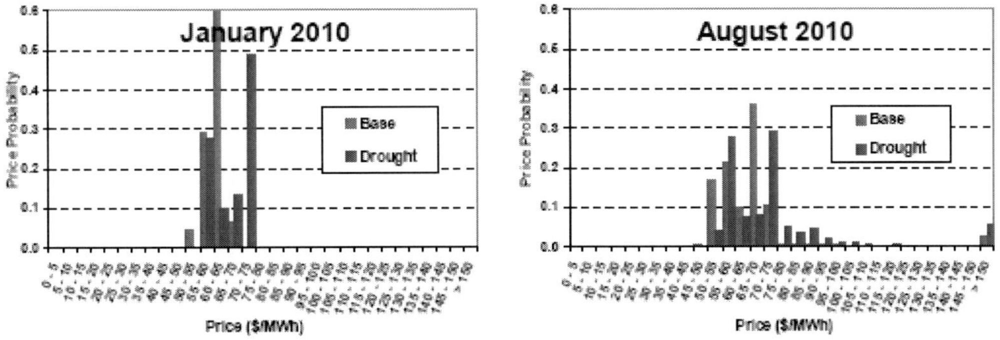

Figure 25. Price Distribution for January and August 2010.

3.3.4. Impacts on CO_2 Emissions

Emissions of CO_2 were calculated for each scenario. The results are listed in Table 7. In the drought scenario, CO_2 emissions were higher by about 20 million tons in each year simulated. On a percentage basis, the increase was rather small; emissions were 5.4% higher in 2010 and fell to 4.3% higher in 2015, and 3.8% higher in 2020.

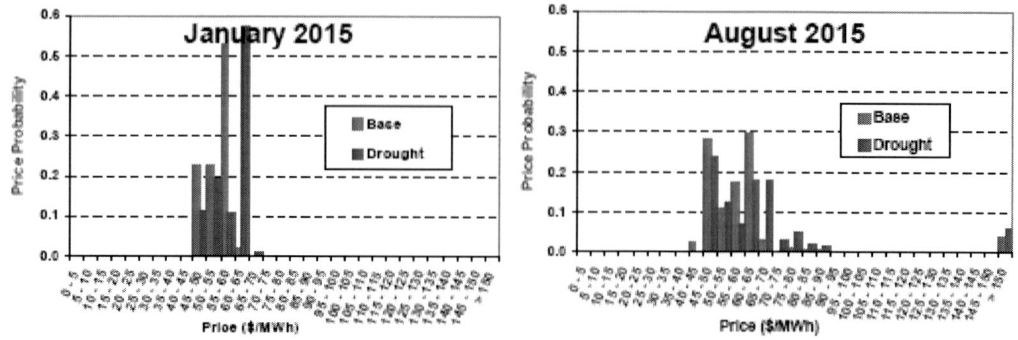

Figure 26. Price Distribution for January and August 2015.

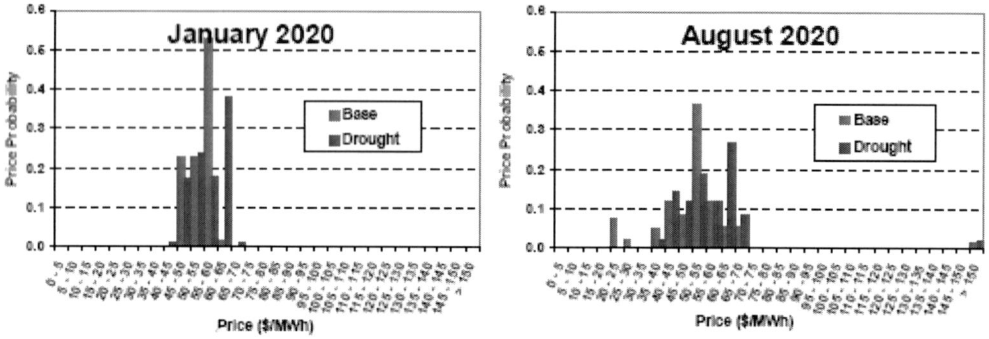

Figure 27. Price Distribution for January and August 2020.

Table 7. Comparison of CO_2 Emissions — Base and Drought Scenarios

Year	Base Scenario (106 tons of CO_2)	Drought Scenario (106 tons of CO_2)	Difference (106 tons of CO_2)
2010	408.4	430.5	22.1
2015	480.5	501.3	20.8
2020	548.1	569.1	21.0

Because natural gas-fired power plants generate the vast majority of electricity to replace the capacity lost by the shutdown of coal plants and the reduction in generation by hydropower plants in the drought scenario, the increase in CO_2 emissions m ay not be as high as expected.

This could be because (1) natural gas generates less CO_2 per Btu than coal, and (2) the natural gas plants that would have produced the electricity for the shut-down coal plants are slightly more efficient than coal plants (i.e., they have a lower heat rate or use less fuel to produce a unit of electricity).

4. Summary and Conclusion

This study resulted in a number of important observations regarding the operation of the electric power system in the western United States and how system operation changes caused by severe drought conditions, particularly in the near term (i.e., less than 10 years in the future). This is the time period when utilities would have difficulty bringing a sufficient amount of new capacity online in response to persistent drought conditions, other than those plants already in the construction pipeline. These observations can also be applied to electric power systems in other parts of the United States to provide some insights into how they might be affected during a drought.

4.1. Effect of Drought on Generation Mix

One observation is that natural gas plants replaced virtually all of the generation lost as a result of plant shutdowns. In the WECC regions, more than 94% of plants that draw fresh surface water for cooling use coal for fuel, while fewer than 6% of those plants use natural gas. Natural gas plants were in the best position to make up for the lost generation because they are operated at much lower capacity factors than coal plants. The average capacity factor of natural gas plants in the WE CC regions is less than 40%, while the average capacity factor of coal plants exceeded 80%. Therefore, the natural gas plants have excess capability to produce more electricity.

Our study showed that other sources, such as nuclear and renewables, are unable to provide more electricity for various reasons. Nuclear power plant growth is constrained, and they are already operating at their maximum capacity factors. Renewables, such as wind, geothermal, and hydroelectric, are already maximizing their energy capacity.

This observation could be applied to power systems in other parts of the United States. Coal-fired power plants are very prevalent in all U.S. power systems, and they typically operate at very high capacity factors because of their low operating cost. They also use large quantities of water, much of which is supplied from fresh surface water sources. Therefore, a heavy reliance on natural gas plants is likely in the near term to replace power lost to plant shutdowns as a result of drought. Natural gas plants operate at moderate capacity factors, between 25% and 50%, and therefore have the capability to produce more electricity quickly.

Electric power systems in the United States that do not have sufficient natural gas plant capacity to replace electricity lost by plant shutdowns that result from drought would have a difficult time generating the needed energy, particularly in the near term. For example, in North Carolina, only 2.5% of electricity generation comes from natural gas, while 60% comes from coal, 32% from nuclear, and 3.5% from hydro (Vinluan 2007). With this type of generation mix, providers may have a difficult time meeting their customers' electricity needs during a drought; they may have to purchase power on the open market at prices that are likely driven up by drought conditions.

However, this study shows that systems that rely heavily on coal plants would realize significant benefits in the long term by building new coal plants equipped with advanced cooling technologies to reduce their vulnerability to drought conditions.

4.2. Effect of Drought on Energy Prices and Water Supplies

Electricity costs in the drought scenario are very high compared with costs in the base scenario in the first 5 to 10 years, but the cost difference grows smaller with time. This is because new coal plants come online steadily and begin generating more of the electricity lost from plant shutdowns that result from drought. The new coal plants use advanced cooling technologies, such as dry cooling, that are much less vulnerable to drought conditions. Coal plants can be three times less expensive to operate as natural gas plants.

With natural gas plants picking up the slack for other plants shut down during a drought, use o natural gas by power generators will increase, which will likely raise the price of natural gas in the market. Because natural gas is used domestically for cooking and heating, consumers may see not only their electric rates increase, but also their domestic natural gas rates.

This has already been happening in the last several years; power generators have beenconstructing natural gas electric plants because they can be constructed more cheaply than large coal plants and can come online faster because they are smaller and face less opposition by the local population. However, quantification of natural gas price impacts is outside the scope of this study.

Some utilities have already recognized the drought problem and have taken action to diversify their water supplies. In 2004, the owners of the Laramie River Station in Wyoming negotiated rights to purchase groundwater from local landowners and installed a 90,000-foot-long pipeline to deliver groundwater to supplement cooling water from the Grayrocks Reservoir. Pipeline operation began in October 2004 (Heartland Consumer Power District 2005).

Produced water from a coal bed natural gas project in the Powder River Basin has been propose as a source of cooling water for both the Laramie River and Dave Johnston power stations (All Consulting 2006). Also, in 2004, the Nebraska Public Power District spent $12 million and installed 40 wells at its 1,300-MW, coal-fired Gerald Gentleman Station to ensure there will be enough water in the event that Lake McConaughy goes dry (Laukaitis 2004).

Diversification of water supplies, particularly for large steam turbine power plants such as coal and nuclear plants, will have to be seriously considered in other parts of the United States. Droughts have already presented a problem in the southeastern United States (in 2007); such problems are likely to continue in the Southeast and may affect other regions in the future. However, groundwater use may not be aviable solution in all cases because groundwater is often used for other, more important purposes, such as for drinking water.

4.3. Effect of Drought on CO_2 Emissions

Increases in CO_2 emissions from changes in electric power system operations that occur due to a drought appear to be minor. In this case study, CO_2 emissions increased just over 5% in the drought scenario compared with the base scenario. Although natural gas plants increased their generation dramatically, the higher efficiency of these plants, coupled with a CO_2 emission factor for natural gas that is almost half that of coal, meant only slightly increased CO_2 emissions. Similar results would be expected for other U.S. electric power systems if their proportion of coal to natural gas generation is similar to that in the West.

4.4. Effect of Drought on Use of Nuclear Power

Drought could have a serious effect on nuclear power plants, in addition to coal plants. In this case study, the cooling systems of the nuclear power plants located in the WECC regions predominately used cooling sources other than fresh surface water, such as ocean water and sewage effluent. Plants with these cooling sources are much less likely to be shut down or have capacity curtailments during a severe drought. However, other parts of the United States rely more heavily on nuclear power plants that receive cooling water from fresh surface water sources.

As recently as summer 2007, the southeast region of the United States faced a very severe drought, prompting North Carolina to develop contingency plans to manage power plant output in response to falling water levels (Vinluan 2007). Power systems in the United States that rely heavily on coal and nuclear power should be studied more carefully to evaluate their vulnerability to drought and to determine whether mitigation strategies are needed.

4.5. Areas for Future Study

This study did not account for transmission constraints, which may curtail delivery of electric power from the generating station to the load. We assumed that any generator in the system could send electricity to any load. In effect, the spatial component of loads and generators was not taken into account. Under normal operating circumstances, this is a reasonable assumption; however, in some circumstances, this may oversimplify the problem and not yield reliable results. Studying the effects of a drought may be one of those circumstances because droughts can affect a very specific area without affecting other areas.

In reality, the transmission system can impose severe constraints on transferring power from one area to another. Transmission lines in some areas may be insufficient to handle normal loads, let alone heavy loads. Also, some transmission lines may be sufficient under normal operating conditions, but could easily become overloaded under extreme circumstances. Environmental conditions, such as excessive heat which often accompanies a drought, can also limit the electric capacity of transmission lines.

Many areas of the United States have transmission corridors in which the lines are very close to their operating limit; severe circumstances can easily overload those lines. This study could be enhanced to (1) account for constraints in transmission capacity and (2) evaluate how that may affect power plant operations during drought conditions. The WECC system includes several transmission corridors in which transmission lines have serious power transfer constraints, particularly lines that serve high-population centers like Los Angeles. There are transmission bottlenecks in many other parts of the United States as well because system loads and the generating capacity serving those loads can be concentrated in areas far apart.

This study focused on plant shutdowns or curtailments due to low water intake levels caused by droughts. However, droughts o ften occur with very hot conditions, which may result in other effects not taken into account in this study. Power plants have limits on the temperature of water they return to the cooling source. Most plants cannot discharge water warmer than 90° to 110° If the water temperature is too high, the plant must curtail the power level so that the water delivered back to the source is below the threshold value. This condition often occurs in July and August — the months of peak load, not only in the WECC regions, but also in most of the United States. Plants that are not affected by low water levels may be affected by temperature limits for cooling water discharge. This condition could curtail more capacity in an electric power system already affected by drought and further exacerbate the problem. This study could be enhanced to take this effect into account.

Also, plants that may not use any cooling water could be affected by excessive heat, because if intake air is too hot, plant power output is reduc ed. This problem, which can occur in gas turbines in hot summer months, is often remedied by humidifying the inlet air. This study could be enhanced to evaluate the extent of this issue and determine whether it may substantially affect model results.

References

All Consulting, 2006, Feasibility Study of Expanded Coal Bed Natural Gas Produced Water Management Alternatives in the Wyoming Portion of the Powder River Basin, Phase II, report prepared for U.S. Department of Energy, National Energy Technology Laboratory, Wyoming State Planning Office, Feb.

Cramton, P., and J. Lien, Value of Lost Load, University of Maryland, Feb. 14, available at [www.isone.com/committees/comm_wkgrps/inactive/rsvsrmoc_wkgrp/Literature_Survey _Value _of_Lost_Load.rtf], accessed Feb. 5, 2009.

EIA: See Energy Information Administration.

Energy Information Administration, 2006, The Electricity Market Modu le of the National Energy Modeling System Model Documentation Report, DOE/EIA-M068, Office of Integrated Analysis and Forecasting, U.S. Department of Energy March, available at [http://tonto.eia.doe.gov/ftproot/ modeldoc/m068(2006).pdf], accessed Feb. 2009.

Energy Information Administration, 2008[a], Annual Energy Outlook 2008 with Projections to 2030, June, available at [http://www.eia.doe.gov/oiaf/archive/aeo08/index.html], accessed Feb. 2009.

Energy Information Administration, 2008[b], Assumptions to the Annual Energy Outlook 2008: Electricity Market Module, Report Number DOE/EIA-0554(2008), U.S. Department of Energy, June, available at [http://www.eia.doe.gov/oiaf/aeo/ assumption/ pdf/electricity.pdf], accessed Feb. 2009.

Energy Information Administration, undated[a], Form EIA-860 — Annual Electric Generator Report; Form EIA-767 — Steam-Electric Plant Operation and Design Report; Form EIA-423 — Monthly Cost and Quality of Fuels for Electric Plants Report; and Form EIA-906 — Power Plant Report, available at [http://www.eia.doe.gov/oss/forms.html], accessed Feb. 2009.

Energy Information Administration, undated[b], Voluntary Reporting of Greenhouse Gases Program, available at [http://www.eia.doe.gov/oiaf/1605/coefficients.html], accessed Feb. 2009.

Federal Energy Regulatory Commission, 2009, FERC Form 714 — Annual Electric Balancing Authority Area and Planning Area Report, updated Jan. 8, available at [http://www.ferc.gov/ docs-filing/eforms/form-714/overview.asp], accessed Feb. 2009.

FERC: See Federal Energy Regulatory Commission.

Heartland Consumers Power District, 2005, "The Power," Heartland Consumers Power District Newsletter, June.

Kimmell, T.A., and J.A. Veil, 2009, Impact of Drought on Cooling Water Intakes, DOE/NETL2009/1364, prepared for the U.S. Department of Energy, National Energy Technology Laboratory, April, 91 pp.

Laukaitis, A.J., 2004, "Five Years of Drought Have Cost Billions," Nebraska Journal Star, Oct. 11, available at [http://www.journalstar.com/articles/ 2004/10/11/local/ doc416a0da 331bbc 338650190.txt], accessed Feb. 6, 2009.

NERC: See North American Electric Reliability Corporation.

North American Electric Reliability Corporation, 2008, Generating Availability Data System (GADS), available at [http://www.nerc.com/page.php?cid=4%7C43], accessed Feb. 2006.

University of Nebraska Lincoln, 2009, U.S. Drought Monitor, Feb. 3, available at [http://drought.unl.edu/DM/MONITOR.HTML], accessed Feb. 2009.

WECC: See Western Electricity Coordinating Council.

Western Electricity Coordinating Council, 2007[a], WECC 2007 Power Supply Assessment, May 9, available at [http://www.wecc.biz/documents/library/ publications/Power Assessment/2006_Power_Supply_Assessment.pdf], accessed Feb. 2009.

Western Electricity Coordinating Council, 2007[b], Summary of Estimated Loads and Resources, June.

Vinluan, F., 2007, "Drought Could Force Shutdown of Nuclear, Coal Plants," Triangle Business Journal, Nov. 23, [available at http://triangle.bizjournals.com/triangle/ stories/2007/11/26/ story5.html], accessed Feb. 6, 2009.

In: Dams, Drought and Energy-Water Interdependencies
Editor: Daniel L. Calzi
ISBN: 978-1-62808-942-4
© 2013 Nova Science Publishers, Inc.

Chapter 3

ENERGY AND WATER SECTOR POLICY STRATEGIES FOR DROUGHT MITIGATION*

Andjelka Kelic, Verne Loose, Vanessa Vargas and Eric Vugrin

ABSTRACT

Tensions between the energy and water sectors occur when demand for electric power is high and water supply levels are low. There are several regions of the country, such as the western and southwestern states, where the confluence of energy and water is always strained due to population growth. However, for much of the country, this tension occurs at particular times of year (e.g., summer) or when a region is suffering from drought conditions. This report discusses prior work on the interdependencies between energy and water. It identifies the types of power plants that are most likely to be susceptible to water shortages, the regions of the country where this is most likely to occur, and policy options that can be applied in both the energy and water sectors to address the issue. The policy options are designed to be applied in the near term, applicable to all areas of the country, and to ease the tension between the energy and water sectors by addressing peak power demand or decreased water supply.

ACRONYMS AND ABBREVIATIONS

Acronym	Definition
^{o}F	degrees Fahrenheit
AIC	administratively induced cost
CCR	Code of California Regulations
CEER	Center for Energy and Environmental Resources
DOE	U.S. Department of Energy
DSM	demand-side management
DWP	Department of Water and Power

* This is an edited, reformatted and augmented version of Sandia National Laboratories, Publication No. SAND 2009-1360, dated March 2009.

EIA	Energy Information Administration
EPRI	Electric Power Research Institute
FY	fiscal year
IOU	investor-owned utility
IPP	independent power producer
MW	megawatt(s)
MWh	megawatt-hour
NCDC	National Climactic Data Center
NGCC	natural gas-fired, combined-cycle
PHDI	Palmer Hydrological Drought Index
PIC	policy-induced cost
PURPA	Public Utility Regulatory Policy Act
RFW	Recycled Water Facility (Watsonville)
SNL	Sandia National Laboratories
U.S.C.	United States Code

1. BACKGROUND

Tensions between the energy and water sectors occur when demand for electric power is high and water supply levels are low. There are several regions of the country, such as the western and southwestern states, where the confluence of energy and water is always strained due to population growth. However, for much of the country, this tension occurs at particular times of year (e.g., summer) or when a region is suffering from drought conditions. In response to the concern of population growth straining resources (such as water) that are used in the generation of electric power, the chairmen and ranking members of the House and Senate Subcommittees on Energy and Water Development Appropriations issued a letter to the Secretary of Energy on December 9, 2004, requesting "a report on energy and water interdependencies, focusing on threats to national energy production that might result from limited water supplies" (Visclosky et al. 2004).

In 2007, the U.S. Department of Energy (DOE) issued the report "Energy Demands on Water Resources: Report to Congress on the Interdependency of Energy and Water" (DOE 2007). A consortium of DOE national laboratories and the Electric Power Research Institute (EPRI), under the direction of DOE, developed the congressional report that highlights many of the issues and challenges that have been identified by energy and water officials and managers across the country. DOE, in cooperation with several national laboratories, is also currently finalizing another report that provides an "Energy-Water Research Roadmap" (Sandia National Laboratories [SNL] 2007). The roadmap discusses research and development efforts necessary to reduce water used by energy production.

Technological solutions to the problem, as described in the roadmap, will take years to achieve and implement. Drought conditions are occurring now and will continue to occur. The study described in this report identifies the types of power plants that are most likely to be susceptible to water shortages, the regions of the country where this is most likely to occur, and policy options that can be applied in both the energy and water sectors to address the issue. The policy options are designed to be applied in the near term, applicable to all areas of the country, and to ease the tension between the energy and water sectors by addressing peak power demand or decreased water supply.

2. Previous Studies on Energy and Water

Because the work conducted in this report builds off of prior efforts on energy and water, this section provides a general overview of "Energy Demands on Water Resources: Report to Congress on the Interdependency of Energy and Water" (DOE 2007) and the "Energy-Water Research Roadmap" (SNL 2007). Unless otherwise stated, all data and conclusions presented in this section are taken from those reports. The reports provide a general discussion of water use in the electric power industry, along with conclusions and future directions that serve as a background for understanding the policy options that are presented in this report.

2.1. Energy Demands on Water Resources

In response to Congress's request for an assessment of energy and water interdependencies, the DOE submitted "Energy Demands on Water Resources: Report to Congress on the Interdependency of Energy and Water" on January 17, 2007 (DOE 2007). The congressional report provides statistics indicating that water resources are already limited (Figure 2-1) and that the available resources have the potential to become even more scarce if the U. S. population grows, as projected, and current trends and policies in energy and water use efficiency do not change. Given this context, the congressional report focuses on three key areas:

- Energy and water interdependencies,
- Water shortage impacts on energy infrastructure, and
- Technologies and policies to make better use of water resources in the context of energy production.

2.1.1. Energy and Water Interdependencies

Water Uses for Energy Extraction and Production

The energy sector uses water resources in four major categories:

- Electric power generation,
- Energy extraction and fuel production,
- Refining and processing, and
- Energy transportation and storage.

DOE (2007) discusses both the quantities of water used for power generation processes and the impacts of those processes on water quality.

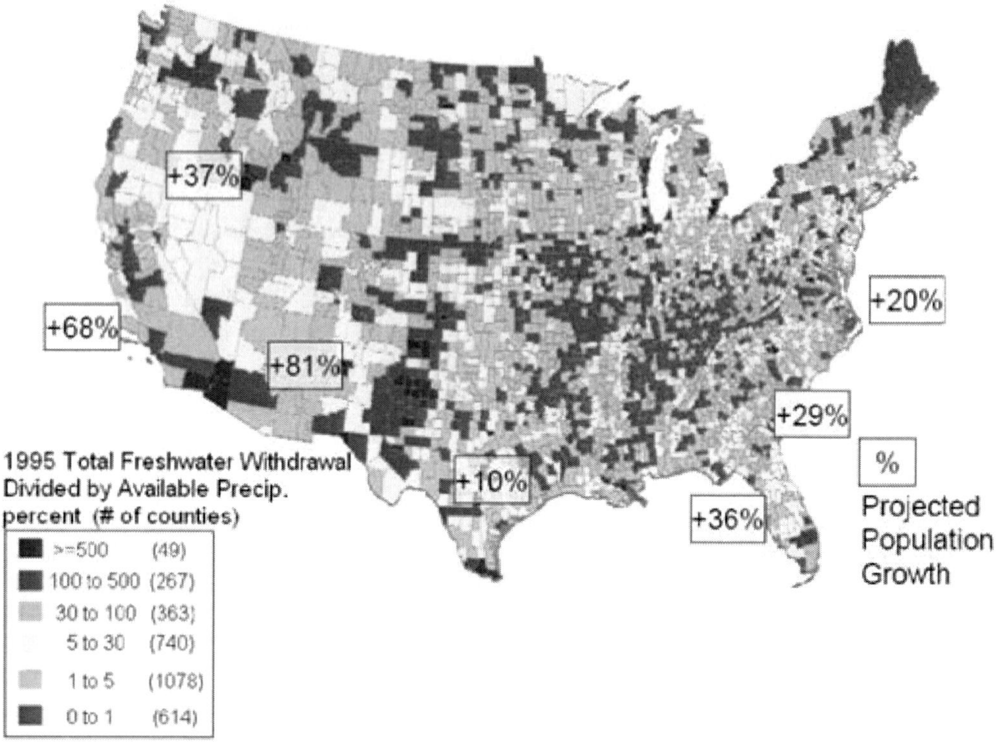

Figure 2-1. Water Shortages and Population Growth (Figure I-2 in DOE 2007).

Thermoelectric power generation uses surface and groundwater for cooling and scrubbing. The quantities of water used and consumed[1] in these processes are dependent upon the type of cooling system. For example, open-loop cooling systems use much larger quantities of water than do closed-loop cooling systems. Open-loop systems lose only a small percentage of the entire volume to evaporation, while closed-loop systems lose most of the withdrawn water to evaporation. Consequently, open-loop systems withdraw more water than closed-loop systems, but both systems actually consume similar amounts. Water is also used in hydroelectric power generation. There are many different types of hydroelectric power generation plants, but the largest source of water loss from this process occurs from water evaporation when hydroelectric power generation involves large reservoirs. The congressional report further states that surface waters and ecology are affected by the thermal and air emissions from both thermoelectric and hydroelectric power generation processes.

The congressional report also discusses the many different ways that water is used for energy extraction and fuel production. For example, water is required in the growing of feedstocks for biofuels, and the quantity of water consumed per gallon of fuel produced can be very high. Alternatively, drilling and mining industries use water to cool and lubricate drilling equipment when drilling for natural gas and oil. Consumption of water in this process is not as intensive as it is in the production of biofuels; however, drilling regulations require appropriate treatment and disposal of the water to minimize environmental impacts. Water is also required to refine and process oil, gas, and other fuels. These processes can result in wastewater that must be treated. The congressional report lists many other uses of water in the energy-extraction and fuel-production processes and indicates that the methods in which water is used greatly affect both the quantity and the quality of the water.

Finally, the congressional report details many different methods that use water for the transportation and storage of energy resources. For example, coal is often transported by barges on rivers, and the movement of barges through locks can affect water levels. Water is also used for transportation in the hydrostatic testing of oil and gas pipelines. Before a new pipeline can be used to transport natural gas or oil, it must be tested by pumping pressurized water through it. This process results in wastewater requiring treatment. Additional uses of water for transportation and storage of energy resources are listed in the congressional report.

Energy Required for Supplying Water

The congressional report indicates that the waste water treatment and water supply sectors are significant consumers of electric power. In fact, the EPRI reports that water treatment and distribution consumes approximately 4 percent of all electricity generated in the United States (EPRI 2002a). The cost of electricity accounts for approximately 75 percent of the total cost of municipal water processing and distribution (Powicki 2002). As population growth continues, limited water supplies will require that water be transported from farther distances and extracted from greater depths. Transportation over greater distances will result in a higher energy requirement per unit volume of water. Additionally, more stringent water requirements may emerge, increasing energy consumption.

2.1.2. Impacts of Water Shortages

The congressional report provides a brief, introductory discussion of how water shortages could affect energy production. Private citizens and public officials are becoming more concerned and aware of the potential ramifications that power generation can have on water supplies. Instances have been documented in which power plants have had to limit power production due to water limitations.[2] In some cases, the development of new power plants has been opposed because of the impacts they would have upon water supplies. In short, limitations on water supplies are leading to conflicts between water managers and electric power production. The potential consequence of water-imposed power generation restrictions is a less stable and less reliable power grid.

The congressional report further analyzes how future power generation could affect water supplies. Projected thermoelectric power plant retirements and additions (Hoffman et al. 2004) were analyzed to assess future water requirements by thermoelectric power generation. While water withdrawals are not expected to change significantly over the next 20 years, water consumption could increase drastically if evaporative closed-loop cooling systems are used for new and replacement plants (Figure 2-2).

2.1.3. Technologies to Increase Water Efficiency in Power Generation

The congressional report identifies a number of technologies that could be employed to increase water usage efficiency in the process of generating power. Table 2-1 provides a summary of the technologies and pros and cons associated with each of the technologies. The congressional report notes that these technologies will likely not be employed until they are economically feasible, and the feasibility will be determined by the scarcity of water resources.

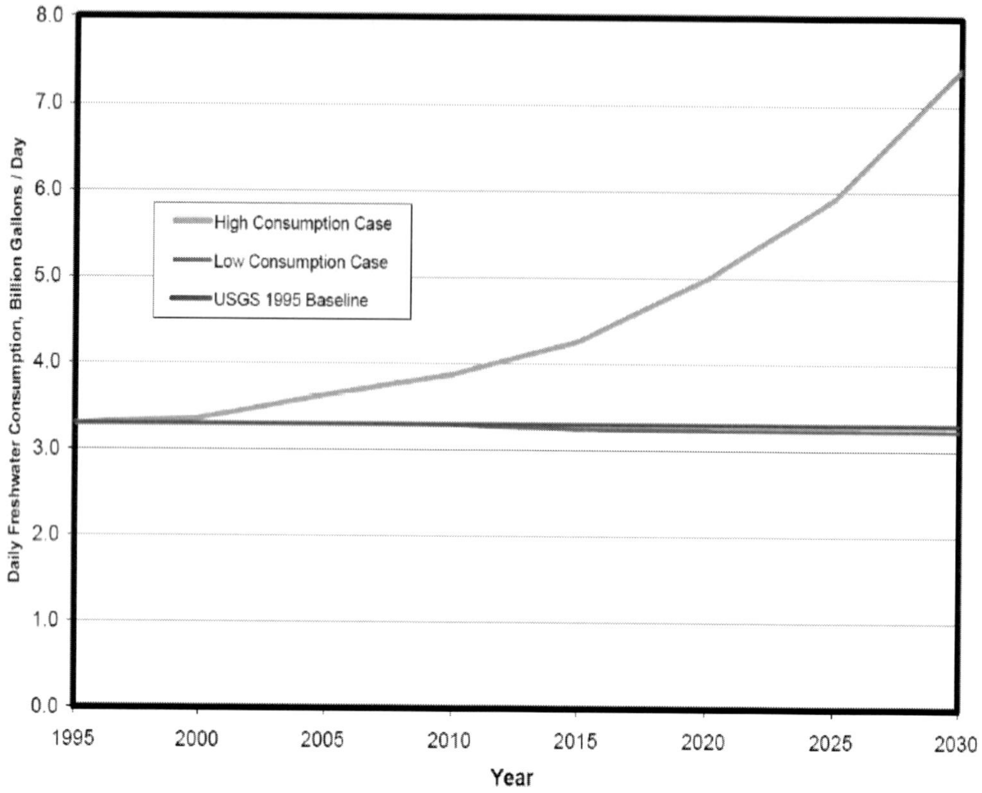

Note: "High Consumption Case" refers to the scenario in which closed-loop cooling systems are used for new and replacement plants.

Figure 2-2. Projected Freshwater Consumption for Thermoelectric Power Generation (Figure IV-7 in DOE 2007).

Each of the technologies has drawbacks and must overcome significant hurdles before being implemented on a large scale in the energy sector. The congressional report concludes that the technologies' drawbacks must be addressed by a complete evaluation of how water policies affect energy supplies and demands and how energy policies affect water supplies and demands. To do this, the following steps are recommended:

- Collaboration on water and energy resource planning among federal, regional, and state agencies with industry and other stake holders;
- Evaluation of natural resource policies and regulations to determine potential, unintended consequences on energy and water sectors(science-based, system-level approaches can be used to advise policy makers); and
- Coordinated development of energy and water infrastructures to reduce conflicts between the two sectors.

Table 2-1. Technologies for Improving Water Use Efficiency in Power Generation

Technology	Examples	Pros	Cons
Advanced cooling for thermoelectric power plants	Dry (air) cooling, hybrid (wet and dry) cooling	Reduces water use	Cost, complexity, performance in hot weather, scalability to large plants
Combined-cycle gas turbines	Natural-gas-fired, combined-cycle gas turbines, integrated gasification, combined-cycle power plants	Reduces water use by half	High cost of gas and increased dependence in gas imports
Renewable electric power	Wind, solar, run-of-river hydroelectric, ocean energy systems	Reduces water use, provides peak power needs, carbon free	Cost, manufacturing/ deployment capacity, need for storage at high penetration (for some technologies)
Oil shale	Recovering oil from oil shale deposits	Large domestic supplies	Cost, potential water demand from extraction process, technology required to mitigate environmental impacts
Renewable and alternative fuels	Biofuels, synfuels, hydrogen	Renewable, carbon-neutral domestic fuels and fuels from domestic coal and gas	Technology development required, cost, high water use for current biofuel production techniques
Increasing and stretching water supplies	Use of degraded water, coordinated energy, and water conservation	Improve water supply understanding and usage	Lack of data, water storage needs to be increased, climate variability, new policies required, coordination needed

2.2. The Energy-Water Research Roadmap

Beginning in fiscal year (FY) 2005, Congress began to provide funding for the development of an energy-water science and technology research roadmap. A roadmap is defined as

> ...a strategic plan to identify and implement the research and development needed to address technical and programmatic issues and challenges associated within a specific area from a system-level context in order to maximize the valued outputs, such as cost effectiveness, reliability, security, and sustainability, and improve overall performance and support public welfare. (SNL 2007)

A consortium of DOE national laboratories, led by SNL, has applied this concept to the energy-water resource issue, and a report documenting their work is currently being finalized. Because the roadmap is not yet complete, details of the analysis are not presented in this section; however, the section does include a general discussion of the methodology for the analysis and the resulting conclusions.

A series of regional workshops was held to identify and assess major regional and national issues, challenges, and concerns associated with energy and water needs and development. More than 500 participants from a spectrum of sectors attended the workshops, including representatives from energy and water utilities and industries, water and energy managers, regulatory agencies, environmental groups, researchers, and tribal organizations. Based upon the information gathered from these workshops, four recommendations were made concerning science and technology research and development efforts. They are

- Reduce water use in electric power generation,
- Reduce water use in alternative fuels production,
- Use non-traditional water sources in electric power generation processes, and
- Initiate further integrated resources planning and management.

If these recommendations are implemented, the roadmap asserts that future energy reliability, sustainability, and cost-effectiveness will be improved by reducing water demands from the energy sector.

3. Factors Affecting Drought Resiliency

A goal of this analysis was to determine what characteristics make a given type of power plant resilient to water shortages. We have concluded that a plant's resiliency is primarily determined by two factors:

- The quantity of water required for plant operation: The quantity of water required for plant operation significantly affects a plant's resiliency because plants that need large volumes of water are likely to be affected by water shortages before plants that require lesser amounts of water.[3] The manner in which the plant uses water determines the volume required for plant operation.
- The availability and stability of the plant's water source: Some water sources are rather sensitive to drought, so plants drawing from these sources could be affected by fluctuations in water levels caused by droughts. For example, a plant drawing water from a river is more likely to be affected by drought than a plant drawing water from the ocean.

The following section analyzes how the most common types of electric power plants in the United States use water in the electricity production process. Water use is typically determined by the type of plant in operation, so we will focus on how differences between types of plants affect their water requirements.

3.1. Water Use in Electricity Production

The manner in which an electric power plant uses water is determined by the type of power plant in operation. Thermoelectric and hydroelectric power plants accounted for more than 97 percent of all electricity production in the United States (EIA 2007), so this analysis focuses on how these types of plants use water.

On average, nuclear power plants have the largest capacity per generator (approximately 1,000 megawatts [MW]/generator), followed by coal-fired generators (220 MW/generator), and natural gas-fired generators (80 MW/generator) (Table 3-1). In general, coal-fired and nuclear generators provide the base load electricity and run fairly continuously. Natural gas-fired generators are typically used to provide intermediate load, so they generally are not in continuous operation. The other fuel types tend to provide electricity only during peak load times.

Table 3-1. 2006 Electricity Generating Capacity in Megawatts (EIA 2007)

Energy Source	# of Generators	Generator Nameplate Capacity	Net Summer Capacity	Net Winter Capacity	Net Generation* (Thousand Megawatthours)
Coal	1,493	335,830	312,956	315,163	1,933,723
Petroleum	3,744	64,318	58,097	62,565	55,243
Natural Gas	5,470	442,945	388,294	416,745	617,986
Other Gas	105	2,563	2,256	2,197	33
Nuclear	104	105,585	100,334	101,718	787,218
Hydroelectric (conventional and pumped storage)	4,138	96,988	99,282	98,767	279,689
Other renewables	1,823	26,470	24,113	24,285	61,536
Other	47	976	882	908	6,346
Total	16,294	1,075,677	986,215	1,022,347	3,742,718

*Net generation includes production from utilities and independent power producers.

Figure 3-1 through Figure 3-4 show the geographic distribution of coal-, natural gas-, and petroleum-fired; nuclear; and hydroelectric power plants in the United States. Natural gas- and petroleum-fired power plants are plotted together in Figure 3-2 since some of these facilities can use natural gas and petroleum interchangeably, so the plants are sorted according to their primary fuel sources. Relatively new natural gas-fired plants, however, are designed only to burn natural gas.

Electric power plants primarily use water for one of three purposes:

- Water is used as the primary coolant for thermoelectric power plants,
- Water is the source of energy that is converted to electric power at hydroelectric power plants, or
- Barge transport on waterways is a significant means of transportation for coal, one of the primary energy sources.

Water plays other direct roles in the production of electric power, but the volumes used in those processes are minor in comparison to the quantities required in the three primary processes. Thus, the following subsections discuss how the electric power industry uses water for thermoelectric power plant cooling, hydroelectric power generation, and transportation.

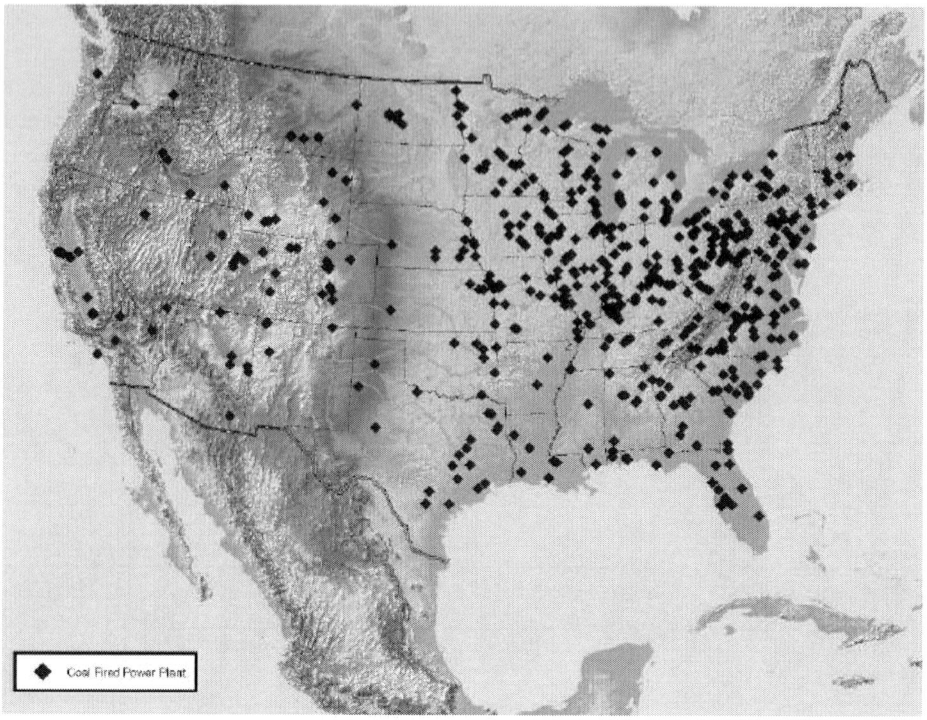

Figure 3-1. Coal-Fired Power Plant Locations in the United States.

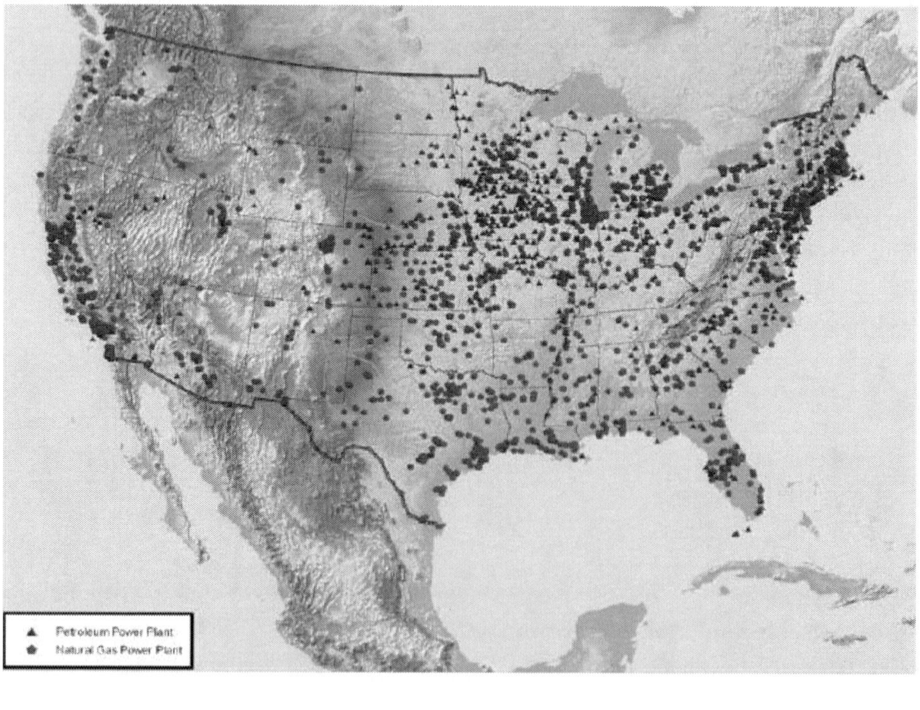

Figure 3-2. Natural Gas- and Petroleum-Fired Power Plant Locations in the United States.

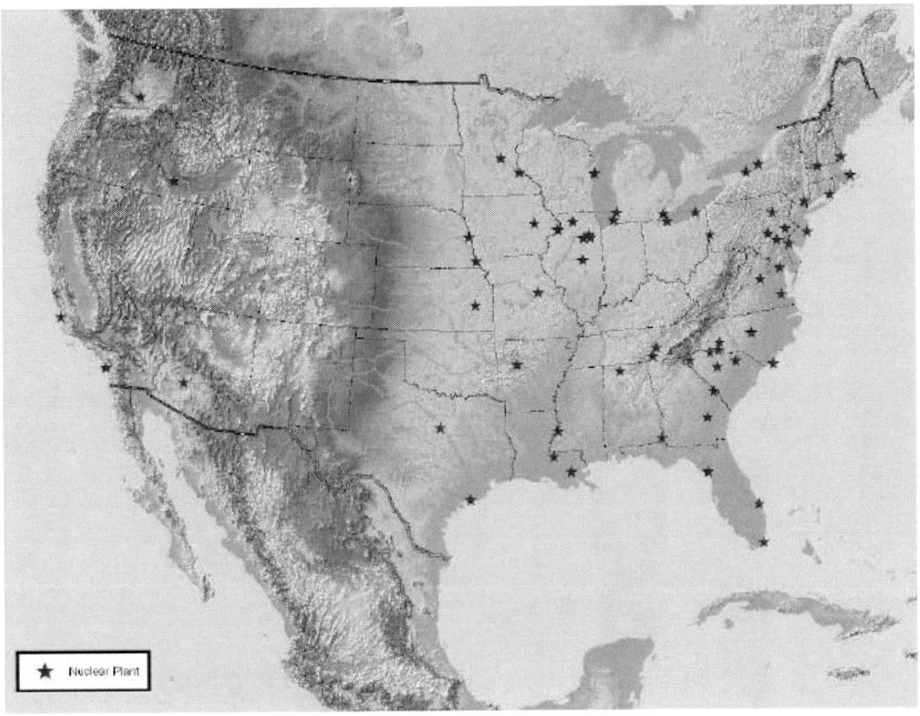

Figure 3-3. Nuclear Power Plant Locations in the United States.

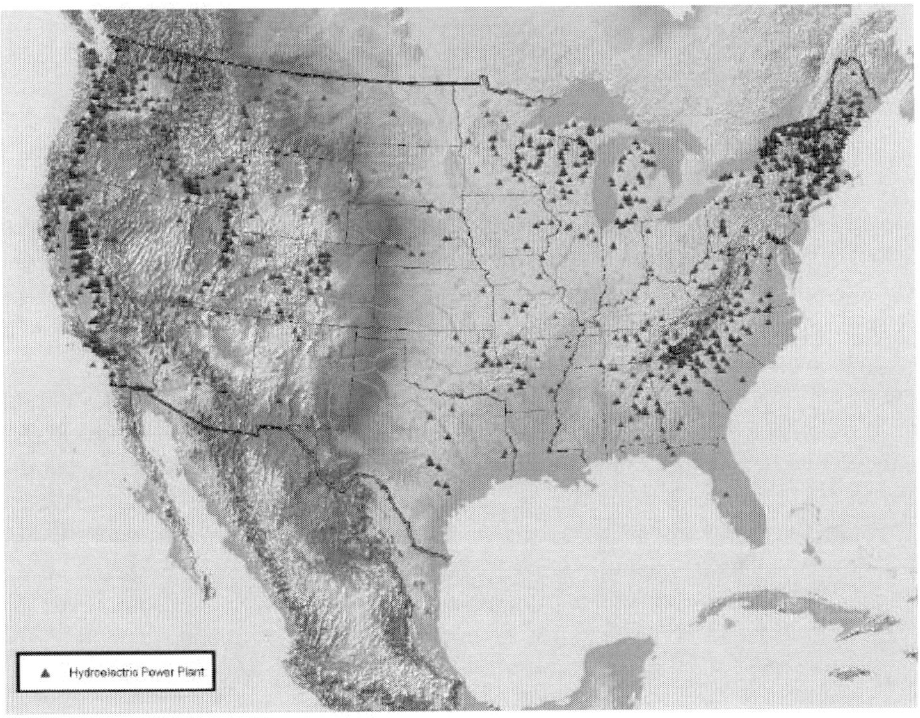

Figure 3-4. Hydroelectric Power Plant Locations in the United States.

3.1.1. Thermoelectric Power Plant Operation

A basic understanding of how thermoelectric power plants operate can illustrate how water is used. Many fossil-fueled (coal, petroleum, etc.) and nuclear power plants generate electricity with a steam turbine. Water is heated in a boiler and becomes pressurized steam. This steam passes through a steam turbine, powering an electric generator. Afterwards, the steam enters a condenser where it is cooled, and the steam returns to its liquid form. The water then circulates back to the boiler, and the entire process is continuously repeated. Water is generally used as the coolant in the condensers. Figure 3-5 illustrates this process.

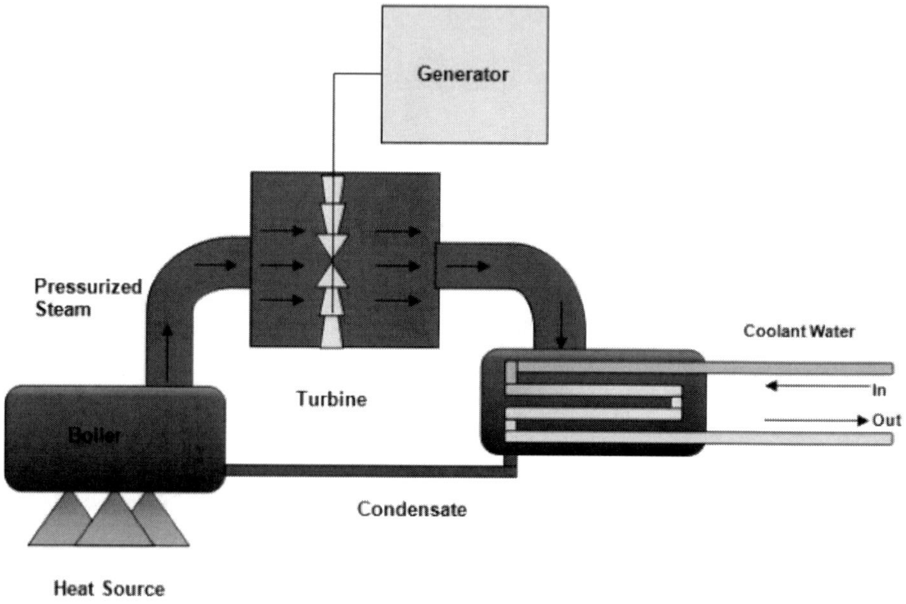

Figure 3-5. Schematic of Conventional Thermoelectric Steam Power Plant Operations.

Most gas-fired and some oil-fired plants often operate in a manner that does not require steam. Rather than burning fuel to heat water into steam, some plants burn a mixture of air and natural gas, thus forming what is termed “combustion gas.” This combustion gas expands through a turbine, causing the turbine to turn and power the electric generator. Gas and internal combustion turbines operate in this manner (Figure 3-6).

One of two things can then be done with the combustion gas after it passes through the turbine. It can either be released, or as is done in natural gas-fired, combined-cycle (NGCC) plants, the excess heat in the combustion gas can be used to heat water into steam. This steam can then be used to power a steam turbine generator, as in conventional steam turbine power plants. This additional step increases the overall efficiency of the generator. Figure 3-7 demonstrates how NGCC plants operate. The water used to turn the steam turbines is circulated in both the steam and the combined-cycle plants, so this quantity of water is preserved.

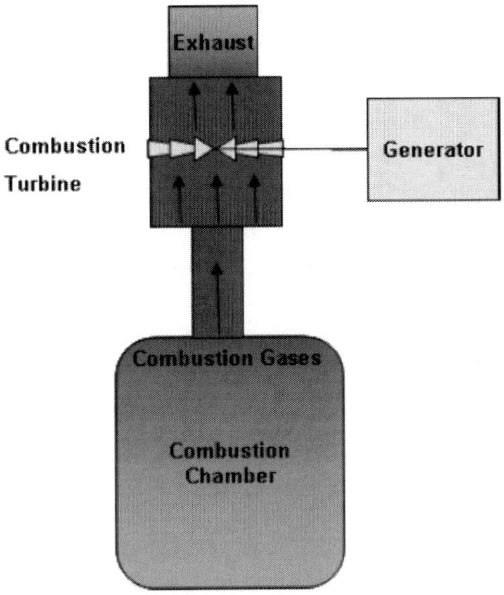

Figure 3-6. Schematic of Combustion Power Plant Operations.

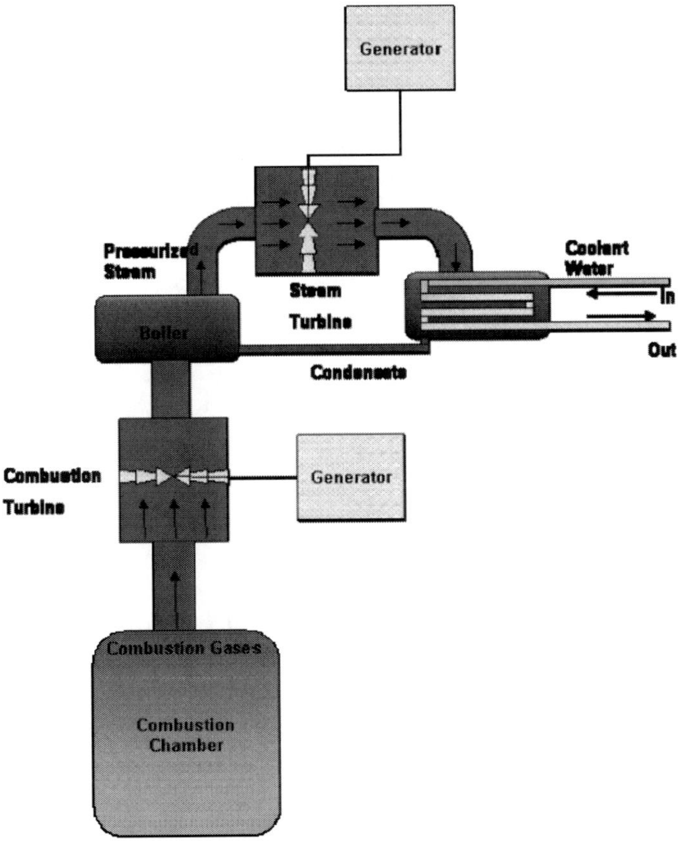

Figure 3-7. Schematic of Combined-Cycle Thermoelectric Power Plant Operations.

In addition to increasing the electricity generation per unit of natural gas, NGCC generators also increase efficiency in terms of MW of electricity generated per gallon of coolant water. NGCC generators derive about two-thirds of their electricity production from the gas turbine and one-third of their production from the steam turbine, so they require approximately a third as much water as fossil-fueled plants for cooling. Thus, NGCC withdrawal and consumption rates are approximately one-third those of fossil-fueled plants (see Section 3.1.3 for further discussion).

Combined-cycle plants have a combined operating capacity of 224,000 MW, comprising approximately 20 percent of the entire U.S. generating capacity. Figure 3-8 shows the geographic distribution of combined-cycle generators in the United States.

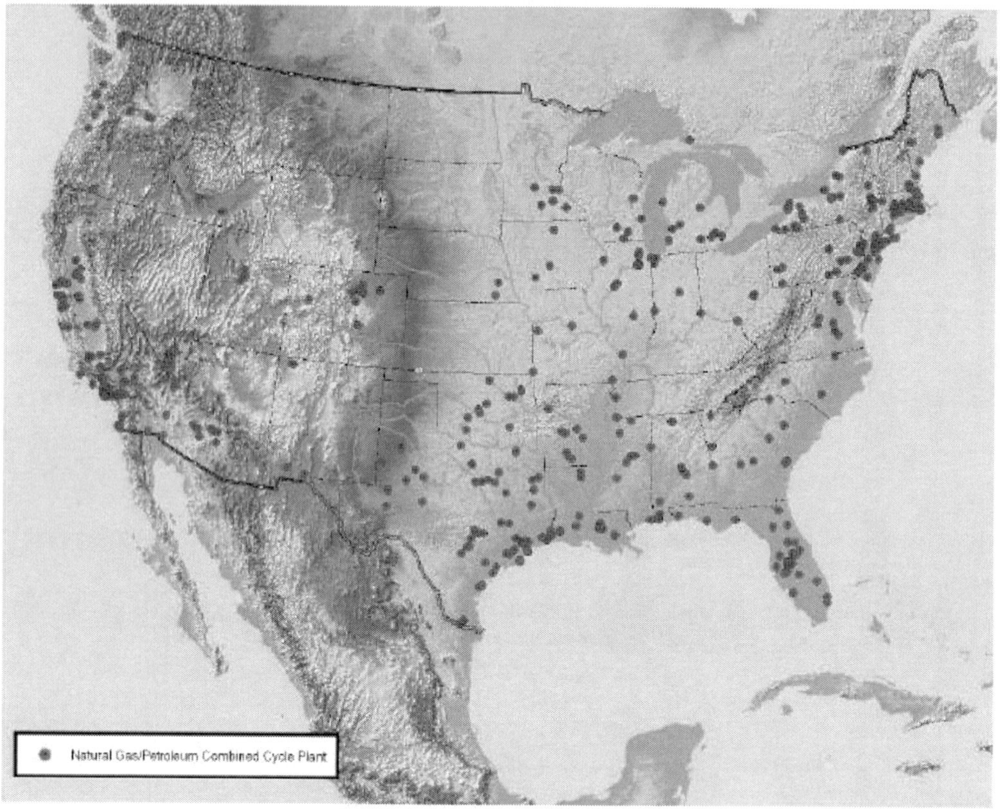

Figure 3-8. Locations of Combined Cycle Plants in the United States.

3.1.2. Cooling Options for Thermoelectric Power Plants

Water is used as the primary coolant in the condensers in both steam plants and combined-cycle plants, and the amount of water used for cooling in these plants is significant. It is estimated that 59 billion gallons of seawater and 136 billion gallons of fresh water are used every day by thermoelectric plants (Hutson, et al. 2004).

Two classes of cooling systems are used. Plants that use "once-through" or "open-loop" cooling systems withdraw large amounts of water from nearby surface water sources, typically a river (Figure 3-9). This water passes through a condenser as the coolant and, in doing so, transfers heat energy from the hot steam to the coolant water, raising the temperature of the water by 12 – 30 degrees Fahrenheit (°F) (EPRI 2002b). After moving through the condenser, the water is released back to the river.

The elevated temperature of the discharged water affects the hydrologic system in several different ways. Increasing the temperature in a river can adversely affect flora and fauna, so the *Federal Water Pollution Control Act* was introduced in 1972 to regulate the impact of open-loop cooling on the environment (33 U.S.C. §§ 1251 – 1387). The increased temperature of the discharge water also increases the rate of evaporation from the river. The quantity of water lost from the hydrologic system by evaporation caused by the elevated temperatures is said to be "consumed."

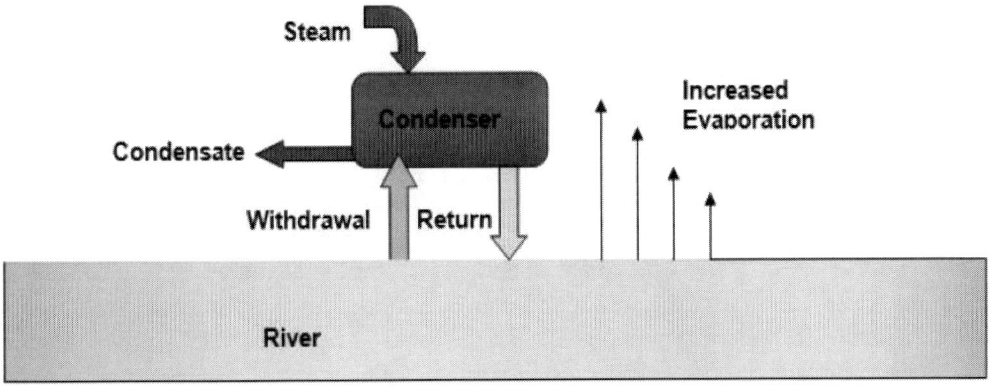

Figure 3-9. Open-loop Cooling System Schematic (adapted from Figure II-2 in DOE 2007).

Prior to 1970, most power plants used open-loop cooling systems, but the enactment of environmental regulations such as the *Federal Water Pollution Control Act* of 1972, Section 316(b) (33 U.S.C. §§ 1251 – 1387), has virtually halted the installation of open-loop cooling systems in new power plants under construction (DOE 2007). Many power plants with open-loop cooling systems remain in operation, but closed-loop cooling systems that have a lesser impact on the environment have become more prevalent in newly constructed power plants.

Closed-loop cooling systems withdraw smaller volumes of water than open-loop systems because they involve a mechanism for circulating a portion of the coolant water. Closed-loop cooling systems that use cooling ponds withdraw cool water from the bottom of the ponds. After the water is passed through the condenser and heated, the water is discharged in a shallow portion of the pond where the heat is allowed to dissipate and the water cools. As happens in the open-loop cooling systems, the addition of heated water to the cooling pond results in some evaporation, and this evaporated volume is considered to be consumed, as well. Eventually, the water cools, its density increases, and it returns to the bottom of the pond where it can be reused for cooling (Figure 3-10).

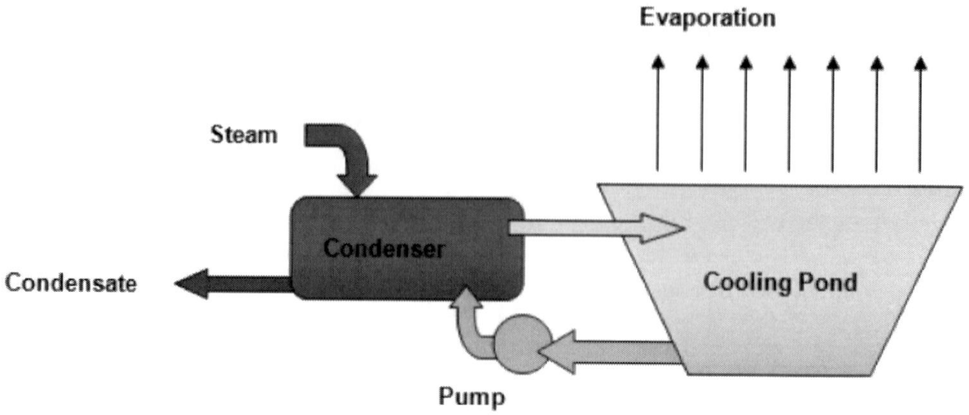

Figure 3-10. Closed-loop Cooling System with a Cooling Pond (adapted from Figure II-3 in DOE 2007).

Some closed-loop systems use cooling towers (Figure 3-11). In wet cooling towers, hot water that is discharged from the condenser is sprayed over metal plates while a fan blows cool air up the tower. The water that evaporates is considered to be consumed, and the remaining water falls down the tower and can be returned to the condenser. Dry cooling towers pump the hot water in small pipes down the tower as fans blow cool air over the pipes. The water eventually cools as it passes through the pipes, and it can be returned to the condenser. No water is lost to evaporation in this process, but dry cooling is not commonly used. Dry cooling relies on the ambient temperature of the air and, thus, can reduce power plant efficiency and output by 25 percent in the summer (DOE 2002). This cooling approach is particularly inefficient in hot, arid climates.

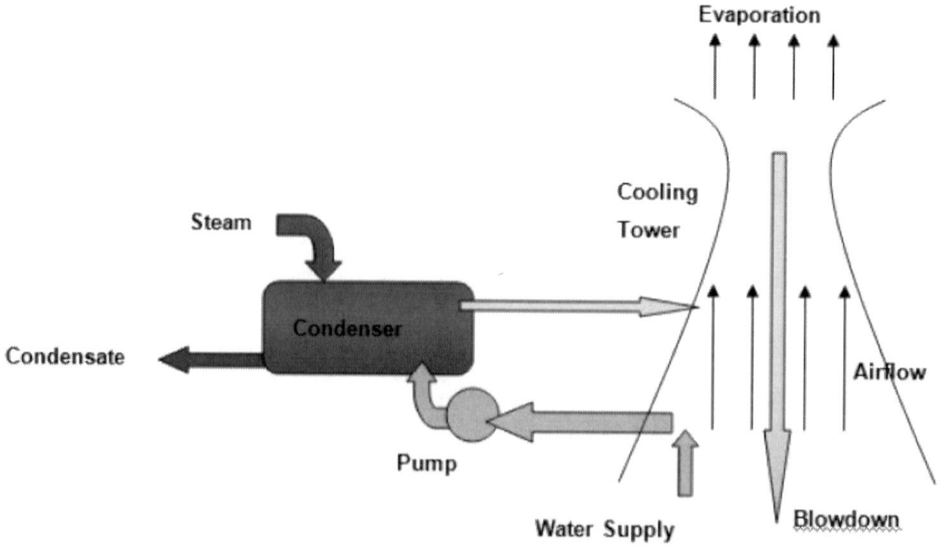

Figure 3-11. Closed-loop Cooling System with Wet Cooling Towers (adapted from II-3 in DOE [2007]).

3.1.3. Water Use Intensity for Thermoelectric Power Generation

Table 3-2 shows the efficiency (in terms of water volume/megawatt-hour (MWh) of electricity generated) of various power plant/cooling system combinations. Because they "employ thermodynamically lower steam conditions than do fossil plants, and thus produce less electricity per pound of circulating steam" (EPRI 2002b), nuclear power plants generally withdraw and consume larger volumes of water per MWh than do fossil-fueled plants and NGCC plants. As previously discussed, NGCC plants derive about two-thirds of their electricity production from the gas turbine and one-third of their production from the steam turbine, so they require approximately a third as much water as do fossil-fueled plants for cooling. Thus, NGCC withdrawal and consumption rates are approximately one-third those of fossil-fueled plants.

Across all plant types, water withdrawals for closed-loop cooling systems are approximately 1 percent of those for open-loop cooling systems. However, consumption levels for open-loop and closed-loop systems (cooling ponds and wet cooling towers) are similar. Dry cooling towers require no withdrawals and do not consume any water.

Table 3-2. Water Usage for Steam Condensing in Thermoelectric Power Plants (Adapted from Table B-1 in DOE 2007)

Plant Type	Cooling Process	Withdrawal (gal/MWh)	Consumption (gal/MWh)
Fossil	Open-loop	20,000-50,000	~300
	Cooling pond	500-600	480
	Wet cooling tower	300-600	300-480
	Dry cooling tower	0	0
Nuclear	Open-loop	25,000-60,000	~400
	Cooling pond	800-1,100	~720
	Wet cooling tower	500-1,100	400-720
	Dry cooling tower	0	0
Natural-Gas, Combined- Cycle	Open-loop	7,500-20,000	100
	Wet cooling tower	~230	~180
	Dry cooling tower	0	0

Source: Table B-1 in DOE 2007, cites EPRI (2002b), CEC (2002), CEC (2006), Leitner (2002) and Cohen (1999).

3.1.4. Hydroelectric Power Plants

Hydroelectric power plants convert the potential and kinetic energy stored in water into electricity. Three different types of hydroelectric power plants have been developed to perform this conversion: impoundment power plants, pumped storage facilities, and diversion facilities.

Impoundment facilities (Figure 3-12) are the most common type of hydroelectric power plant. Typically, a dam is built on a lake or river that has a steep drop in elevation, and the water stored behind the dam is called the "reservoir." The dam raises the water's height and increases its depth, creating water pressure that is termed "head." Water enters the dam through the intake valve and decreases in elevation as it travels through the penstock. A turbine is located at the end of the penstock, and electricity is generated as water passes through the turbine.

Impoundment facilities were typically built by the U.S. Army Corps of Engineers or the Bureau of Land Management and most such facilities were federal projects. Cost/benefit analysis was used to provide the economic justification for such facilities, in part because the facilities were built on federal land. Application of cost/benefit analysis techniques improved over the years such that some benefits (e.g., recreation) that previously were not considered could, by virtue of the improved methods, now be evaluated; many facilities could now be conceived and justified as multi-use facilities, providing a variety of benefits including flood control, electric power, and recreation opportunities. While the flood control and recreation opportunities were considered non-market benefits, electric power was conceived as a market good that would be sold to consumers.

The price at which hydroelectric power was sold typically did not reflect the value of land and the amount of financial outlay provided by the federal government. The only element included in the price of the electricity was the marginal operating cost which, for hydroelectric facilities, is very low. This pricing mechanism has left a lasting impression on the population and on policymakers that hydroelectricity is cheap. This is not, in fact, the case; if the cost of capital, permanent inundation of large swaths of land, and closing off of other recreational opportunities, plus the value of free-flowing rivers, were to be included in the electricity cost, it is likely that hydroelectricity would be among our most expensive generation technologies. For these reasons, dams are no longer being built and there is currently a national discussion regarding the possibility and desirability of removing some of the existing dams.

In addition to these considerations, it has become clear that some of the multi-use benefits may, in fact, be in conflict with one another. For example, in times of drought, it is necessary to spill water to provide for irrigation and for municipal water use, even though it may not be the best time to produce electricity. Also, the level of the impoundment drops in drought conditions, thereby, reducing the hydraulic head and the potential energy that is stored in the reservoir. In fact, during a drought, the "shadow price" of water becomes very high, making hydroelectricity more expensive.

The value of water varies generally between the eastern and western U.S. and between specific water basins, and also between uses of the water. Frederick (1996) states that

> The potential value of water for hydropower within a basin varies widely with the location of the water on the river because the power produced by an acre-foot of water is determined by the developed head (the height of a retained body of water) above the generating turbines. For instance, an acre-foot of water at the headwaters of the Snake River in the Pacific Northwest could pass through 16 dams before joining up with the Columbia River and then through another 4 dams before reaching the Pacific Ocean. The cumulative developed head of these dams is 2,159 feet. In contrast, the developed head of Bonneville Dan, the last dam along the Columbia River, is 59 feet. Consequently, the value for hydropower of an acre-foot of water at the headwaters of the Snake is more than 36 times the value just above Bonneville Dam....Hydropower is an important, although not the highest value, water user in these four water resources regions.

Pumped storage facilities also use dams to store water, and two reservoirs are used (Figure 3-13). One reservoir, located above the dam, has a higher elevation, and the second reservoir is located at a lower elevation below the dam. When there is a high demand for electricity, the pumped storage plant releases water from the upper reservoir to the lower reservoir. As happens in impoundment facilities, the water flows downward through a turbine, and electricity is generated in this process. When the electricity demand is low, the plant pumps water from the lower reservoir to the upper reservoir, and the facility uses some of its own electricity to pump the water.

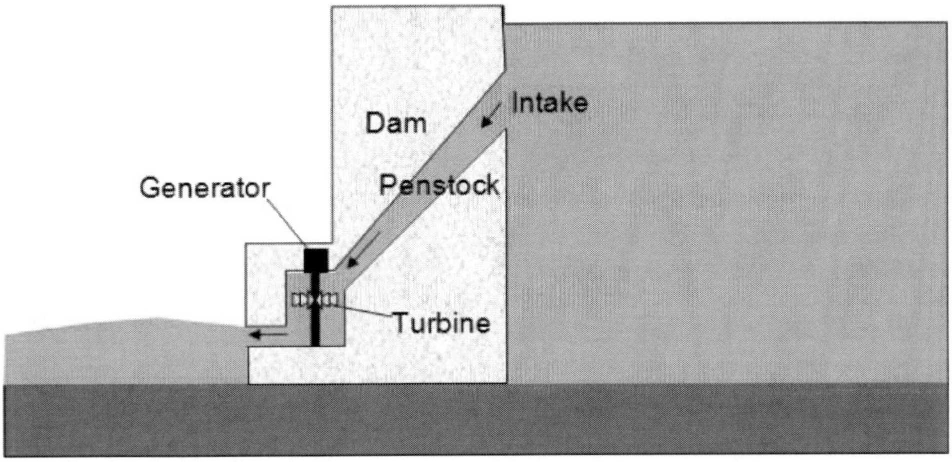

Figure 3-12. Schematic of an Impoundment Hydroelectric Power Plant (adapted from TVA 2008a).

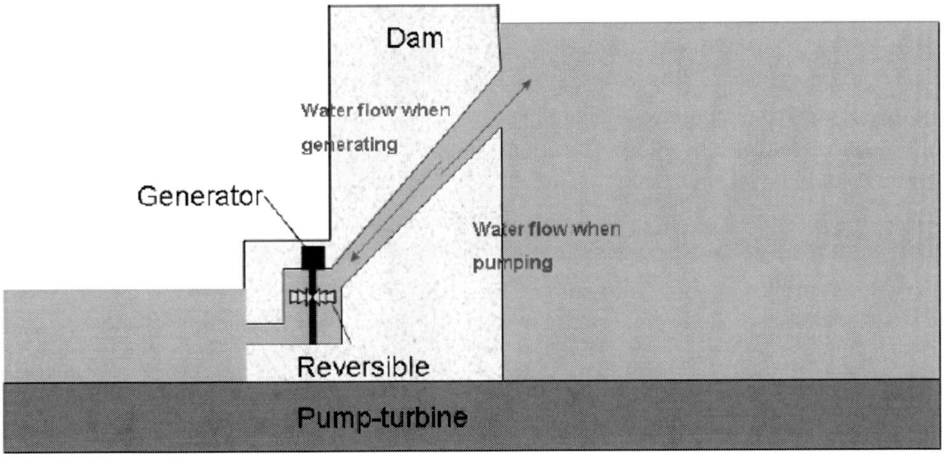

Figure 3-13. Schematic of a Pumped Storage Hydroelectric Power Plant (adapted from TVA 2008b).

None of these plants generate steam or combustion gas, so water is not required as a coolant at hydroelectric plants. Additionally, each of these types of plants requires that sufficient water levels (or depths) must be maintained to generate an appropriate head for electricity production. Electricity production can be restricted or even halted at any of these types of plants if water levels drop too low.

The differences in the designs of these plants determine how their electricity production is affected by water shortages. Once water passes through turbines in impoundment and diversion facilities, it cannot be reused for power generation, so these types of plants are dependent upon precipitation to replenish their sources of water. Consequently, in times of drought when precipitation levels are reduced, water levels can decrease to the point where electricity production is severely limited or halted. For example, in 1988, hydroelectric power generation was reduced by 25 percent along the Mississippi River, following one of the worst U. S. droughts in the past century (Changnon 1989). Furthermore, because water use by municipalities often increases in times of drought, dams may be required to release additional quantities of water, further reducing water levels and hindering power production. Hence, electric power production at impoundment and diversion hydroelectric power plants is sensitive to precipitation levels. In contrast, electricity production at pumped storage facilities is more resistant to drought because the water at these plants can be reused. These plants are more expensive to build, though, and they can be difficult to sight because two reservoirs must be considered.

Diversion, or run-of-river, plants do not require the use of a dam. Rather, the facility diverts a portion of a river through a penstock, into a turbine. The turbine turns as water passes through it, and electricity is generated. Because diversion plants depend entirely on landscape and there is no dam to artificially increase the water's height, diversion plants generally produce a limited amount of power (Schlumberger Limited 2008).

3.1.5. Water as a Means of Transportation

Railroad and barge transport are the primary modes of transporting coal. Transportation of coal by barge on rivers is a critical component of the coal distribution system in the United States. Barge transport is often used to transport coal from the initial source to a railroad, from a railroad to the coal-fired power plant, or the entire distance from the mine to the plant.

Barge transport is typically the cheapest mode of transportation, so when that option is available, it is preferred. However, barge transport and the amount transported on a single barge are dependent upon the depth of the river on which the barge travels. For example, during ideal conditions on the lower Mississippi River, barges sit at a depth of 12 feet below the surface, and barges are lashed together 5 wide and 8 long. In the Fall of 2006, when river depths along the lower Mississippi dropped due to drought conditions, the Lower Mississippi River Committee limited barge depths to 9 1/2 feet and a 4 by 8 lashing configuration (U. S. Water News Online 2006). As a result, the capacity for a string of barges was reduced from 60,000 tons to 28,800 tons. So droughts have the potential to reduce the rate at which all goods, including coal, can be transported by barge.

Some river systems, like the Missouri River, have a system of reservoirs that are used to control river depths. When river levels are low, water is released from the reservoirs to increase river depths and permit barge travel. Additionally, rivers can be dredged if river depths decrease to the point where barge traffic is halted. However, even these river-management techniques may not be sufficient to ensure barge flow continues. An extensive drought beginning in 2006 caused the Missouri River and its aforementioned reservoirs depths to drop to the point where the U.S. Army Corps of Engineers was considering canceling the entire 2008 barge season (Wolken 2006).

Nevertheless, low water levels are unlikely to significantly disrupt electric power generation at coal-burning plants for the following reasons:

- Most coal-burning plants with barge access also can receive coal shipments by rail. Thus, if these plants are unable to receive coal via barge, they will likely be able to receive shipments by rail. During the record-setting drought of 1988, barge traffic was halted twice on the Ohio and lower Mississippi rivers in June and July. Even when barge traffic resumed, it was at a below-average rate (Changnon 1989). Consequently, rail transport was substituted for some of the coal that would normally be transported by barge, but because barge is the cheapest mode of transportation, utilities were paying a higher rate for transportation.
- To offset fluctuations in the ability to receive coal, coal plants generally keep some reserve coal onsite. A rule of thumb is that coal plants keep 1 month's worth of coal onsite. Plants that have access to only a single mode of coal transport likely keep more onsite, perhaps up to 3 months worth in the winter. Plants that have access to multiple modes of coal transport may attempt to keep less onsite because, if one mode of transport is disrupted, they can use the other mode(s) to continue to receive coal.

4. Water Shortage Risks

Water shortage risks to power plants are heavily dependent on not only the way the water is used, but the source of the water itself. For example, groundwater sources (such as aquifers) are not likely to be affected by drought, whereas surface water sources (such as lakes and rivers) are likely to have water restrictions imposed in the event of a drought.

4.1. Drought

For power plants that rely on surface water for their operations, either for cooling or for hydroelectric generation, drought can be a major concern. Drought impacts can range from the need to adjust water release schedules to de-rating or even ceasing the operation of plants.

Drought, from the standpoint of hydrological impacts that include reservoir and groundwater levels, is measured using the Palmer Hydrological Drought Index (PHDI). [4] The PHDI is based on the balance of moisture supply and demand for a given climate division, without including manmade changes such as increased irrigation or new reservoirs. PHDI data are available from the National Climatic Data Center, on a monthly basis dating back to 1895, for the 344 contiguous U.S. climate divisions (excluding Alaska and Hawaii). The data used in this report run from January 1895 through April 2008.[5]

The PHDI and other indices classify drought (or wetness) in degrees of severity ranging from normal to extreme. Table 4-1 shows the potential impacts of moderate, severe, and extreme drought as they pertain to water availability in streams and reservoirs (National Drought Mitigation Center).[6]

Table 4-1. Potential Impacts of Drought Severity

Drought Category	Possible Impacts
Moderate	Streams, reservoirs, or wells low; some water shortages developing or imminent
Severe	Water shortages common; water restrictions imposed
Extreme	Widespread water shortages or restrictions

Figure 4-1 shows the moderate, severe, or extreme drought conditions, as a percent of observations, for the climate. The data range runs from January 1895 through April 2008, with observations collected monthly. The western states, including southern California, northern Nevada, eastern Oregon, and western and central Arizona, see some form of drought or pending drought from 30 percent to 40 percent of the time. Portions of eastern South Dakota, central Pennsylvania, and the northeastern lower peninsula of Michigan also fall into this category.

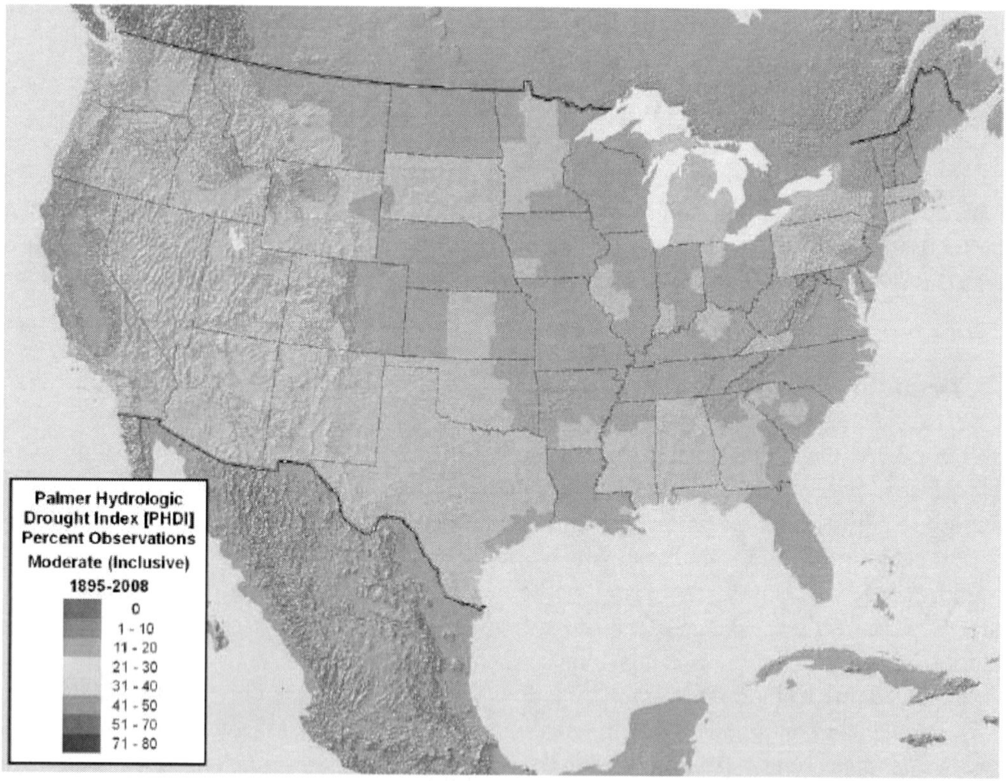

Figure 4-1. Moderate, Severe, and Extreme Drought Frequency by Percent of Observations, 1895-2008.

Because moderate drought does not necessarily result in water shortages, it is more of an indicator that water shortages may occur. Therefore, this analysis focused more closely on instances of severe or worse drought levels. Levels of drought that are considered severe or extreme are likely to require some sort of action from power providers that rely on surface water sources for cooling. Depending on the severity of the drought, those actions could range from different water release timing or waivers for temperature of release to de-rating or even shutting down the plant.

Figure 4-2 shows severe or worse drought conditions for the entire data range from January 1895 through April 2008. As can be seen in this figure, severe or worse drought conditions occur with much less frequency than moderate or worse drought conditions. More frequent occurrences are seen in northeastern Oregon, southwestern Wyoming, and northwestern Arizona. As in the moderate and worse drought conditions, the western states see more frequent occurrences of severe or worse drought than the Midwest or Eastern states.

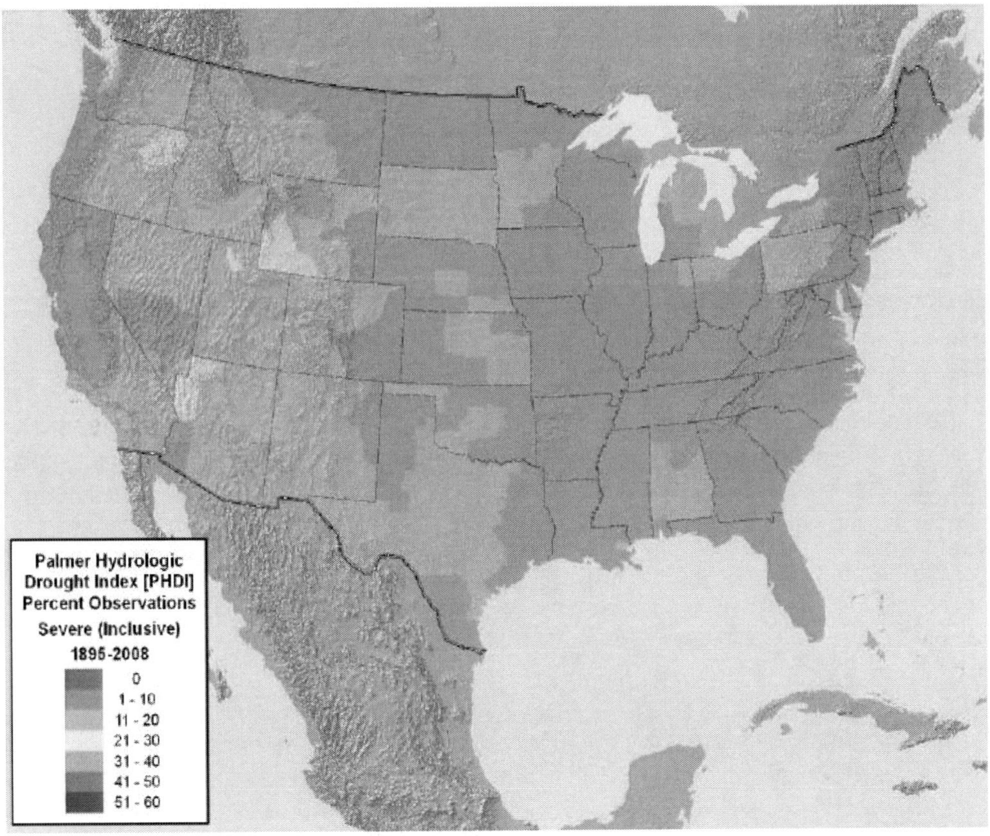

Figure 4-2. Severe and Extreme Drought Frequency by Percent of Observations, 1895-2008.

The data presented in Figure 4-3 through Figure 4-8 are graphed using the same color and frequency scale, with the high point on the color scale being 51 percent to 60 percent of observations as severe or extreme drought. Dividing the data up in 30-year increments allows tracking of drought across the country over time. Figure 4-3 through Figure 4-8 provide this timeline.

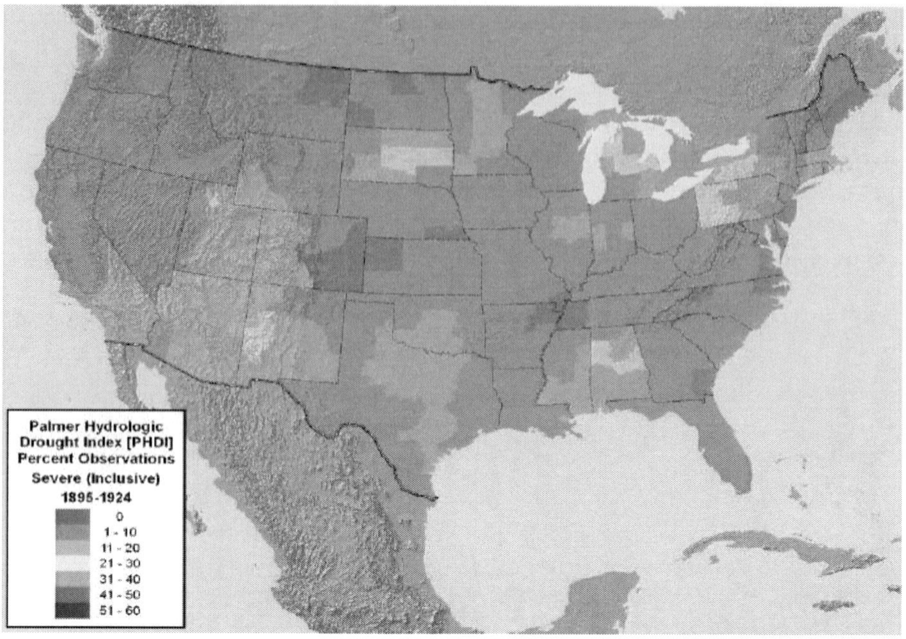

Figure 4-3. Severe and Extreme Drought Frequency by Percent of Observations, 1895-1924.

From the years 1895 to 1924, as seen in Figure 4-3, the most frequent occurrences of severe or extreme drought were in northeastern Michigan, central Pennsylvania, and southern Alabama. This is an unusual set of observations, and drought in those regions does not recur in any of the subsequent 30-year periods.

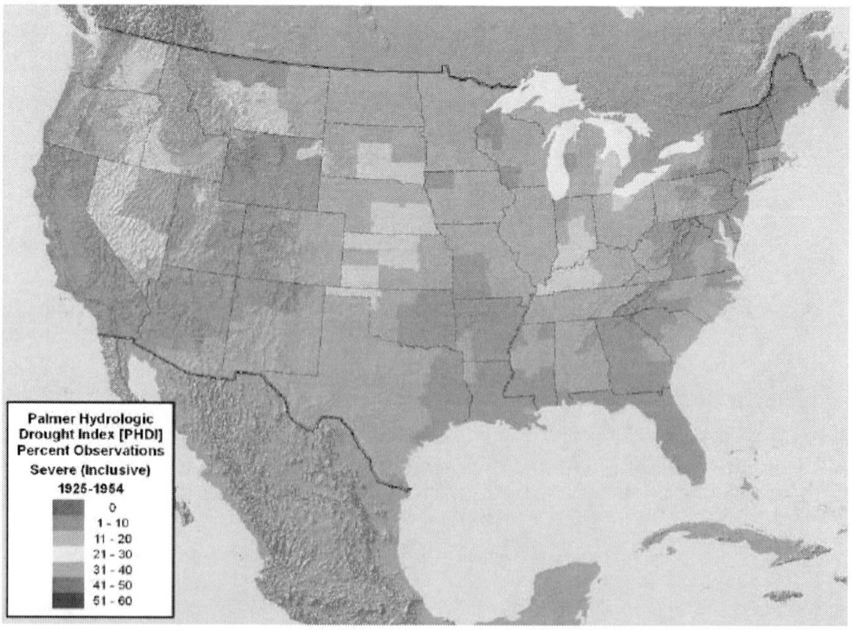

Figure 4-4. Severe and Extreme Drought Frequency by Percent of Observations, 1925-1954.

The years 1925 to 1954, as shown in Figure 4-4, include the dustbowl years of the 1930s with drought in the Great Plains. Drought is also prevalent in the west during this 30-year period.

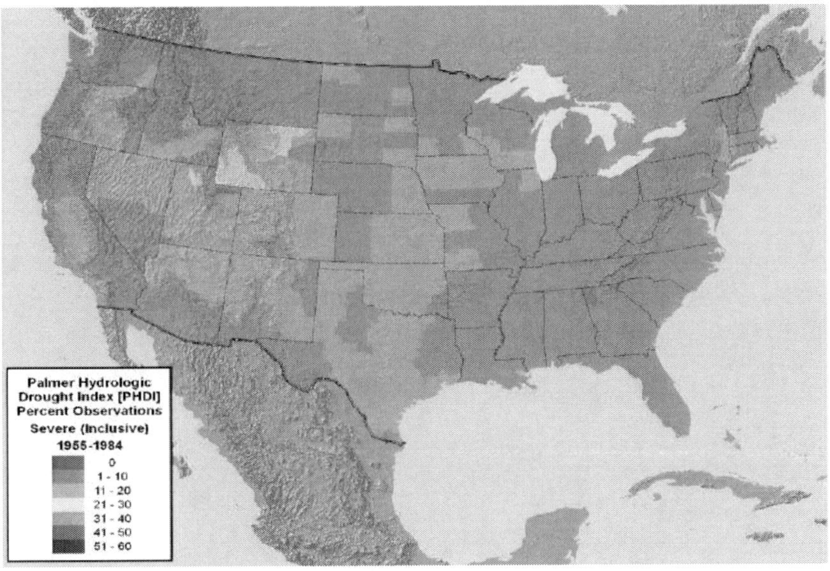

Figure 4-5. Severe and Extreme Drought Frequency by Percent of Observations, 1955-1984.

From the years 1955 to 1984, as shown in Figure 4-5, instances of severe or extreme drought were infrequent, with occasional instances in the west and plains and slightly more frequent instances in Wyoming.

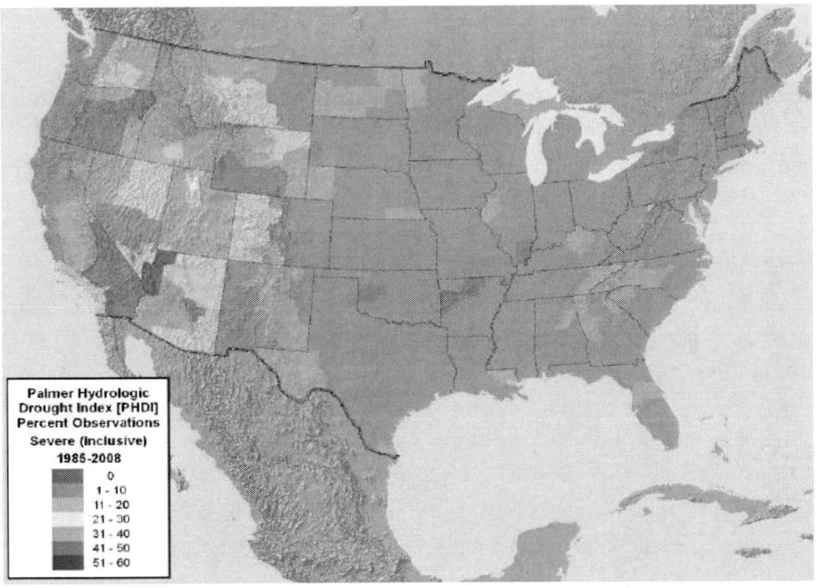

Figure 4-6. Severe and Extreme Drought Frequency by Percent of Observations, 1985-2008.

The final set of data, shown in Figure 4-6, is not a complete 30-year set and spans the remaining 23 years and several months of observations from 1985 to 2008. This period sees a very high frequency of severe or extreme drought in the western states. Even if the remaining 7 years of observations see no drought at all, the percentage of drought occurrences in those states will still be high.

Figure 4-7 and Figure 4-8 examine severe and extreme drought separately for the period 1985 to 2008 to determine what kind of drought is causing the frequent occurrence in the Western states.

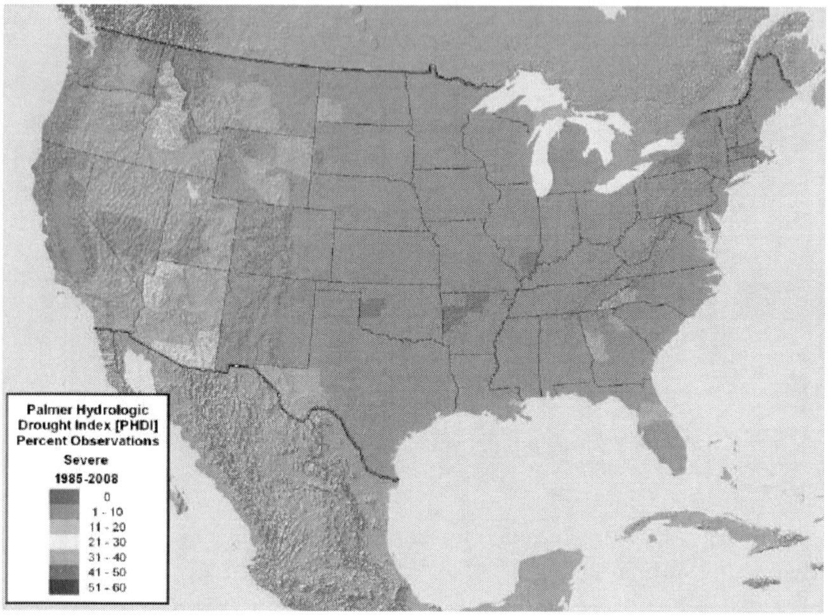

Figure 4-7. Severe Drought Frequency by Percent of Observations, 1985-2008.

As can be seen from these figures, extreme drought is occurring with more frequency than severe drought in the Western states from the period 1985 to 2008. This is an unusual trend compared to the occurrence of extreme drought in the years since data collection began (Appendix A contains the graphs of severe and extreme drought for the 30-year intervals from 1895 to 1984.). If this trend continues, the Western states are likely to see the need for drastic changes in their water-usage patterns. This period is the cause of the high point on the color scale being set to 51 percent to 60 percent of observations. To continue the analysis of the other periods, this period was removed, and the color scale for graphing purposes was recalibrated to a high of 31 percent to 40 percent of the observations being severe or extreme drought.

Figure 4-9, for the period 1895 to 1924, again shows the number of instances of severe and extreme drought. Northeastern Michigan, central Pennsylvania, and southern Alabama still see more frequent occurrences of drought than the rest of the country for this time period. The rescaling of the data also shows that western Pennsylvania, portions of New York, northern Alabama, South Dakota, and New Mexico also see drought about 25 percent of the time during this 30-year period.

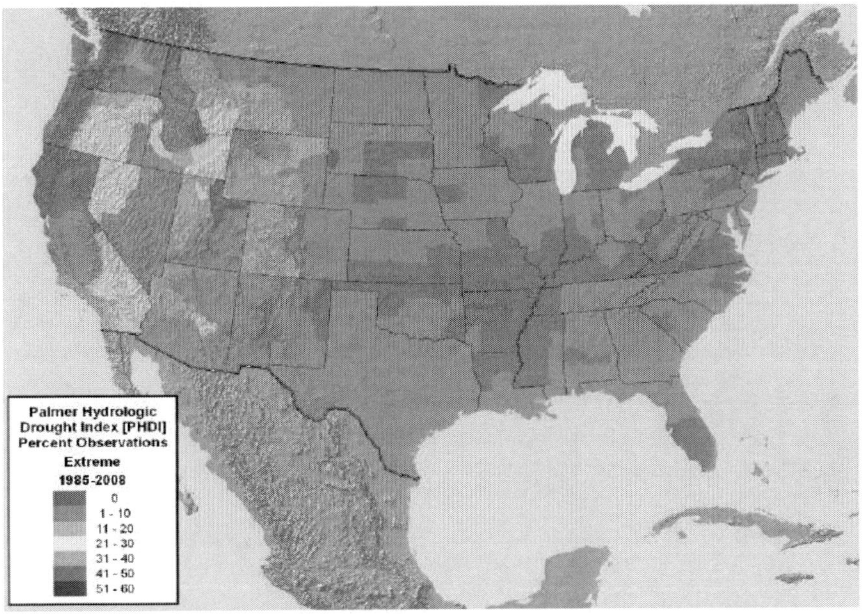

Figure 4-8. Extreme Drought Frequency by Percent of Observations, 1985-2008.

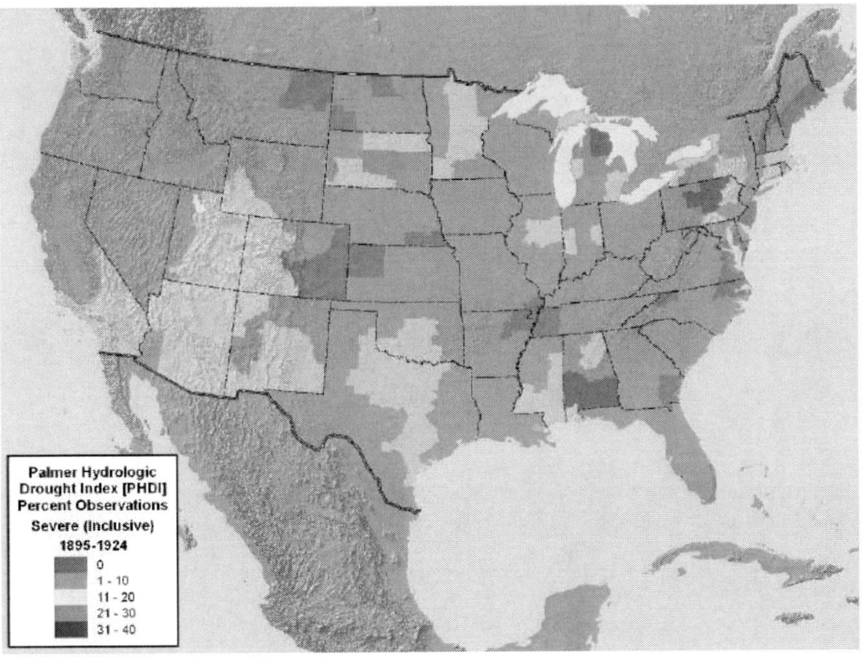

Figure 4-9. Rescaled Severe and Extreme Drought Frequency by Percent of Observations, 1895-1924.

Figure 4-10 covers the period of time that includes the drought and dust bowl of the 1930s. The plains states and the western states see the most frequent occurrences of drought during this timeframe, ranging from 21 percent to 40 percent of the monthly readings.

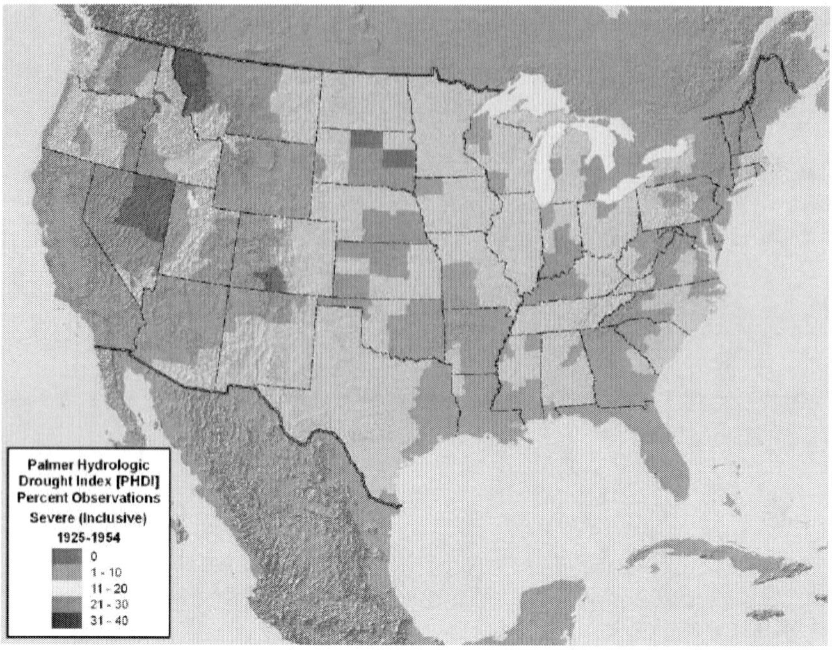

Figure 4-10. Rescaled Severe and Extreme Drought Frequency by Percent of Observations, 1925-1954.

In Figure 4-11, we again see the western states and the plain states with more instances of drought than the rest of the country, but significantly less than in the prior 30-year period.

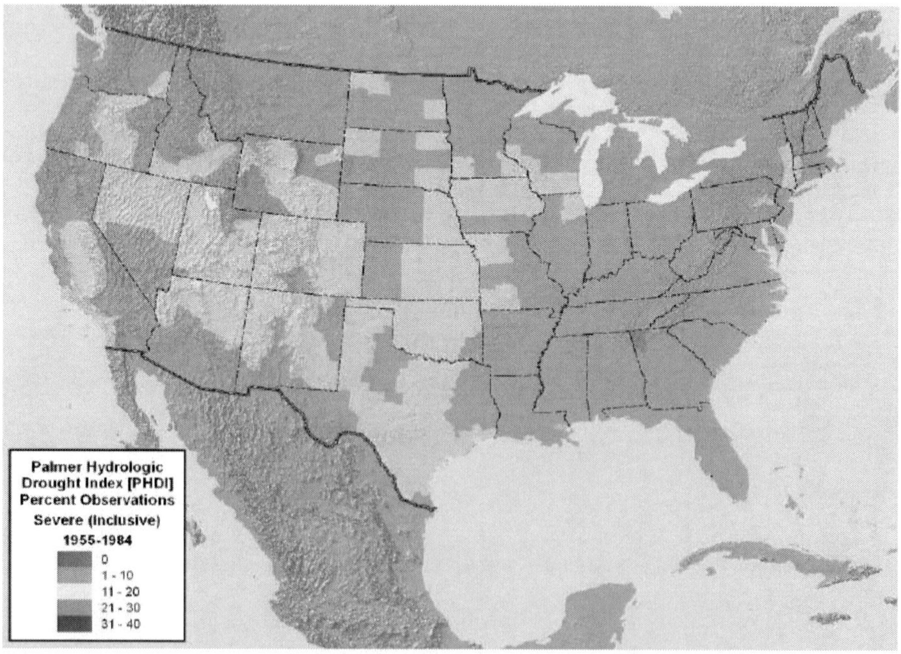

Figure 4-11. Rescaled Severe and Extreme Drought Frequency by Percent of Observations, 1955-1984.

4.1.1. Focus of Effort Based on Drought Analysis

The Western states, in both the most recent data ranges and since data collection began in 1895, have seen more frequent occurrences of drought than the rest of the country. For the period from 1924 through 1955, the Plains states joined their western counterparts and also experienced frequent occurrences of severe and extreme drought.

The southeast saw unusual levels of drought conditions from May 2007 through March 2008, with the Atlanta area seeing its second driest year on record.[7] These conditions do not appear in the graphics due to the duration of the drought (shorter than drought conditions that are endemic to the west and southwest). Also, the metric used, PHDI, only accounts for the moisture balance in the region and not for manmade changes. The method developed in this section, along with the policies that are recommended later in this report can be applied to any region including the southeast. For a discussion of how the recent drought conditions in the southeast compare to those seen in the West, see Appendix A.

The Western region appears to be much more likely to experience and potentially continue to experience conditions of severe or extreme drought. This is also the region with increasing population growth with corresponding increasing demand for energy and water use. This analysis will focus on the Western region to explore possible near-term mitigation strategies for the effects of drought on the energy and water sectors. The next section discusses water management in the Western states as an example to focus the analysis; however the mitigation strategies that are developed can be applied nationwide.

4.2. Water Markets and Water Banks

Preparation for periods of drought is an important problem for urban water service agencies in the western and southwestern United States. Even under normal weather conditions, rapidly growing populations and increasing environmental demands (in-stream flows and wildlife habitats) are putting pressure on local water supplies. Municipal water suppliers have often used water storage and water delivery projects to deal with drought. Both these efforts require capital, energy, and environmentally expensive sites.

The scarcity of water in the western United States led to the development of a doctrine of water allocation termed "prior appropriation," and that doctrine governs water allocation in this area of the country. This doctrine was intended to promote the buying and selling of water in the western United States. In the prior-appropriation doctrine, water rights are not subordinate to the land, as they are in the riparian doctrine and, therefore, can be sold independent of the sale of land. Claims are established by being the first to put water, which has a yearly quantity and appropriation date, to a "beneficial use."

Interest in water markets has recently increased, reflecting growing pressures on limited water supplies, realization that institutional change will be needed to improve water use efficiency, and the shift toward privatization and market mechanisms to address resource-allocation problems. Some proponents of water markets suggest that water markets promote water-use flexibility, establish a recognized water value, and provide incentives for more efficient water use (Gardner 1985). The majority of existing water market literature deals with transactions among agricultural users or between agriculture and urban users.

A water-supply contract is defined as a formal contract or agreement between a farmer or a group of farmers and an urban or industry water user. The contract allows the farmer to transfer water temporarily from agriculture to urban use, during occasional critical drought periods, allowing the urban or industry user to secure a source of drought water supply. The farmer does not necessarily relinquish ownership of the water rights (although the farmer could choose to permanently transfer water rights) and retains access to the water supply during normal water-supply situations.

Water-supply contracts can be similar to stock and commodity exchange market options. In financial terminology this means that the holder of an option contract has the right to buy the commodity (in this case water) at a specified price, termed the strike or exercise price, from the seller of the option. The seller of the option is guaranteeing future delivery under specified conditions and price. In exchange for guaranteeing future delivery of the commodity at a set price, a further premium above the exercise price, called an option price, may be paid to the seller. Water-supply contracts can also take the form of a complete transfer of water rights, temporary transfer of water rights, or rental of a water supply.

Most states operate their water banks at a regional level. Regional banks require fewer resources and are likely to provide an opportunity to identify methods to meet local market requirements. These banks operate mainly as an exchange broker between willing sellers and interested buyers and not as a mechanism by which they can increase in-stream flows.

Market participation is dependent on federal and local policies. Federal policy may prohibit some holders of water rights from participating as sellers. Buyers may be limited to only those with "critical" water needs, as defined by state regulators. Transaction costs can be aggregated into two groups, administratively induced costs (AICs) and policy-induced costs (PICs). AICs include the costs of searching for trading partners and negotiating the terms of the exchange and other contract provisions. PICs include the costs of obtaining approval for water transfers.

4.2.1. Western State Water Markets and Banks

Arizona

The Arizona Water Banking Authority's bank is based on storage credits. It is a market-driven bank in which water rights are leased or purchased, held, sold, and transferred. The bank stores unused Colorado River water and it is either directly recharged into the ground to be held in underground storage facilities or used instead of pumped groundwater.

California

A 6-year drought, from 1987 to 1992, highlighted the tight supply of water across the state of California. In 1991, one of the recommendations of the Governor's Drought Action Team was to establish a state-sponsored water bank operated by the California Department of Water Resources (Hanek 2002). In 1991, inter-basin trades accounted for 101 million cubic meters (m^3) of water, with a value of \$111 million (Howitt et al. 1992). The California water bank also operated in 1992 and 1994 to facilitate water trades during drought and expanded its operations to include a variety of water trades that changed with water availability. In advance of an expected dry season in 1995, the California Department of Water Resources prepared to operate its water bank by purchasing water supply options from sellers. No options were exercised due to increased water supplies.

California's water bank experience has been fairly limited to informal intra-season spot markets and annual lease markets. In California, between 1982 and 1996, only 1.7 million acre feet of water have been transferred. All transfers were spot or short-term (annual lease) transfers (Hanak 2002). No long-term or permanent transfers were arranged by the California water bank; although, since its inception (1982, 1991, 1992, 1993, and 1995), it has handled 40 percent of all of the water transfers that have occurred in the state. In the years the water bank was active, water transfers rarely occurred between agricultural users and commercial or urban users; the majority of transfers were between agricultural users.

In the California water market, AICs and PICs are a function of transaction type as well as property rights. Table 4-2 shows that transaction costs falls asymmetrically between buyers and sellers in the California water market. Heterogeneous property rights and differential rates of market access work to accentuate these asymmetries.

Table 4-2. Transaction Costs by Policy Type

Type of Transaction Cost	Seller	Buyer
Administratively Induced		
Search for trading partners	X	X
Establish price, quantity, and quality	X	X
Negotiate payment terms	X	X
Establish delivery dates	X	X
Negotiate physical transfer	X	X
Policy-Induced		
Identify legal characteristics of water use	X	
Identify hydrological characteristics of rights	X	
Comply with state and federal law of transfer application and approval process	X	
Conduct project approval process	X	
Conduct water district approval process	X	
Adjust costs of changing resource base:	X	
Third-party impacts	X	X
Litigation for damages	X	X
Litigation/risk	X	

Source: Archibald and Renwick 1998a.

The California Department of Water Resources is currently preparing for the establishment of California's 2009 Drought Water Bank (Christie 2008). The bank will buy water primarily from local water agencies and farmers upstream of the San Joaquin-Sacramento River Delta and make it available for sale to public and private water systems expecting to run short of water in 2009. The California Department of Water Resources mandated that agencies buying water through the water bank have to commit to a 20 percent reduction in overall water use (Christie 2008).

Colorado

Colorado has long had active water markets. The data in Table 4-3 show that Colorado, Utah, and New Mexico have had large numbers of permanent water transfers.

Table 4-3. Total Number of Change-of-Water-Right Applications by State, 1975 – 1984

State	Number of Applications Filed
Arizona	30
California	3
Colorado	858
New Mexico	1,133
Utah	3,853
Wyoming	40

Source: MacDonnell 1990.

In addition to permanent sales of water rights, there have been many temporary transfers to accommodate short-term needs, especially in times of drought. During periods of drought, water banks of various types have been organized to facilitate short-term transfers. With regards to permanent water rights transfers, these types of transfers have primarily been from agriculture to municipal and commercial users.

Colorado's water transfers are processed through a water court system, under which proposed transfers must be advertised and can be challenged by parties that perceive themselves to be injured by the transfer. Colorado's reliance on a water court system for administering transfers results in high transaction costs associated with transfers; the transferor bears the burden of the transaction costs.

The establishment of a system to facilitate transfer of water rights, regardless of high transaction costs, has been beneficial to Colorado. It has allowed water to move from lower-valued agricultural uses to higher-valued urban and commercial/industrial uses.

Similar to the California water market, transaction costs fall asymmetrically between sellers and buyers, with sellers/transferors responsible for the majority of transaction costs.

Idaho

Idaho has the longest running water-supply bank, which was authorized in 1979. Idaho has a state-wide water supply bank and three separate rental pools that operate as separate banks. The Idaho Water Resource Board determines the rental rate for the bank and pools to lease water. The three rental pools are situated in watersheds where water is stored in reservoirs and can be released as it is rented. The water-supply bank deals with natural flow water rights and groundwater, as compared with stored water that is rented from the rental pools.

New Mexico

Prior to 2003, in contrast to Colorado and California, applications for new appropriations and for permanent water right transfers were made through the Office of the State Engineer. This office has the technical skills to determine whether or not the new appropriation of the requested transfer will adversely affect other water users. The state engineer's office can then approve or modify the request; buyer and sellers can appeal decisions through the courts if they disagree with the determination of the state engineer. Typically, the determinations of the state engineer are accepted without appeal. The New Mexico system has relatively low transactions costs when compared to the Colorado water transfer system. Legislation in 2003 authorized two pilot water banks, one to be managed by the state engineer and the other by the New Mexico Interstate Stream Commission.

Transaction costs for the buyer or seller/transferor depend on the system each region or state adopts. Regions that rely on courts to facilitate water transfers will incur higher transaction costs than those that operate through state engineers' offices or state-managed water banks. Regardless of the type of organization used to match sellers and buyers of the natural resource, it is likely the seller will incur the largest portion of transaction costs.

5. Assessing the Drought Resilience of Power Plants

As described in Section 3, the resilience of power plants to drought is a combination of two primary factors: the availability and stability of the plant's water supply and the quantity of water required for plant operations. Section 4 assessed the water supply by analyzing drought frequency data to determine the vulnerability of plants to drought, and in this section, that analysis is combined with power plant data to further analyze drought resilience.

Figure 5-1 shows the distribution of power plants around the country, by prime mover,[8] along with the frequency of severe and extreme drought from the period 1985 to 2008. The West and Northeast contain many hydroelectric plants, while the Midwest is dominated by fossil fuel plants. Both of these types of plants are vulnerable to drought conditions. Hydroelectric plants rely on water levels either in reservoirs or run-of-river waters to be able to produce power. In many cases, these plants are used primarily for peaking power or providing additional electrical generation when demand is high. Demand is likely to be high during summer months, which is precisely when drought will be more prevalent, creating a tension between the need to produce additional power and the need to cut back on water usage.

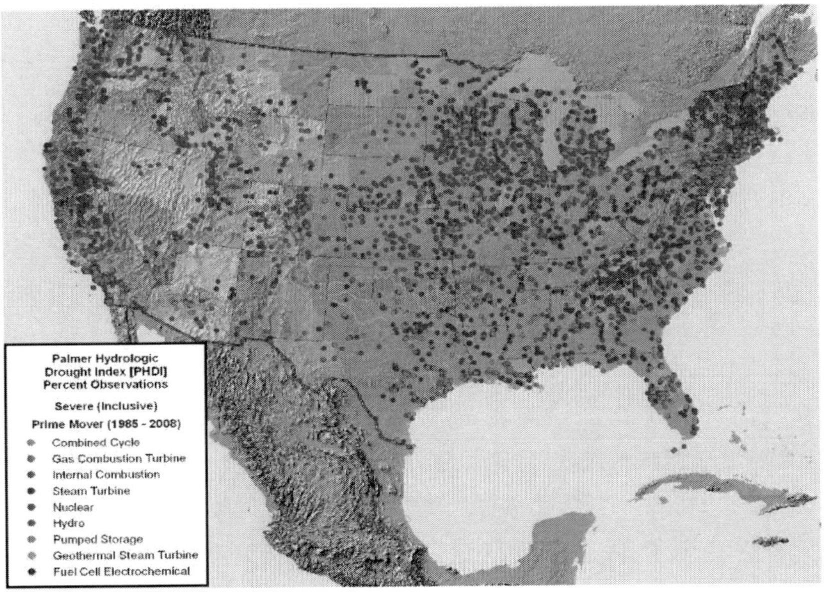

Figure 5-1. Nationwide Power Plants.

Figure 5-2 shows the power plants in regions that suffered from severe or extreme drought in at least 20 percent of the observations from 1985 to 2008. This region contains 784 power plants. Several of the power plants in the region have little to no dependency on water usage, including solar and wind turbine plants. Once these plants are removed from the analysis, 656 plants remain at risk. The details of the types of at-risk plants in region at risk can be found in Table 5-1.

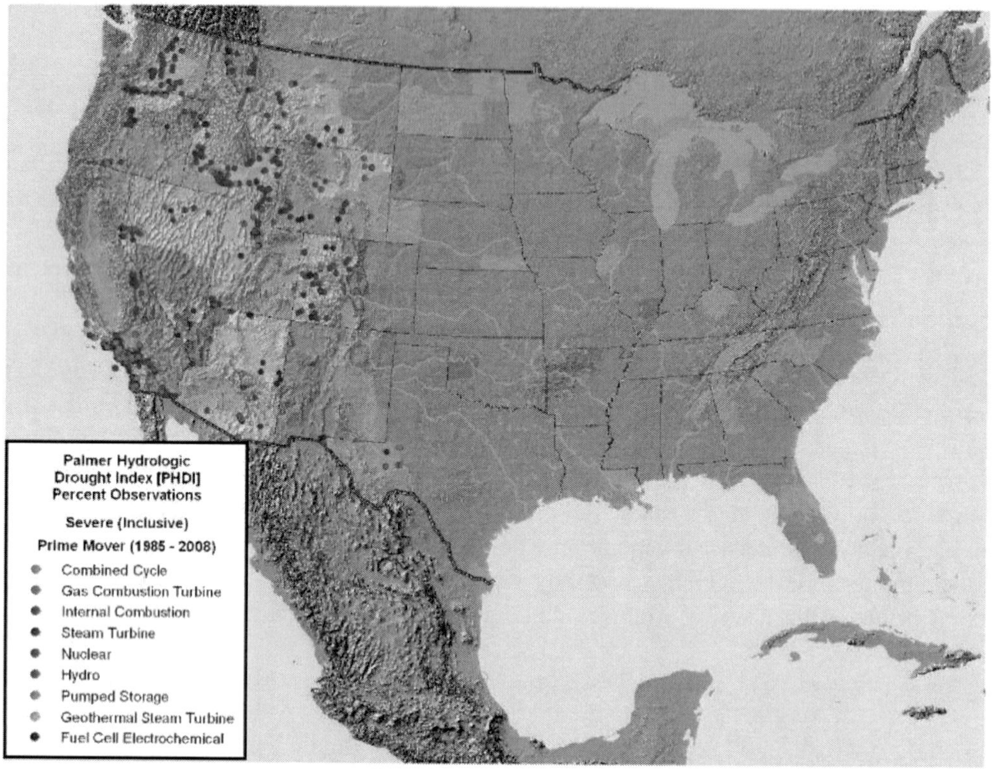

Figure 5-2. Power Plants in Frequent Severe and Extreme Drought Regions, 1985 – 2008.

Table 5-1. Power Plants in Risk Area by Prime Mover and Primary Fuel

Prime Mover	Number of Plants	Primary Fuel
Combined cycle	69	Natural gas
Fuel cell electrochemical	3	Hydrogen
Gas combustion turbine	102	Natural gas
Geothermal steam turbine	44	Geothermal steam
Hydro	279	Water
Internal combustion	75	Natural gas
Nuclear	2	Uranium
Solar	25	Solar
Steam turbine	85	Coal/natural gas
Wind	100	Wind

Additional water usage data, or at least data on cooling mechanisms, are necessary for Sandia to complete the resilience assessment of the power plants. These data are needed to cross link with the plant type and drought information. Sandia was only able to acquire water usage data on a few dozen of the more than 650 plants in high-frequency severe and extreme drought regions; therefore, we are not able to recommend specific plants for more in-depth analysis. With additional data either from DOE or from power plant personnel, Sandia would be able to make recommendations on which specific power plants merit further analysis.

In place of specific plant recommendations, Figure 5-3 contains examples of different types of power plants and their relative resilience to drought. This figure is intended only to be illustrative of possible assessment results.

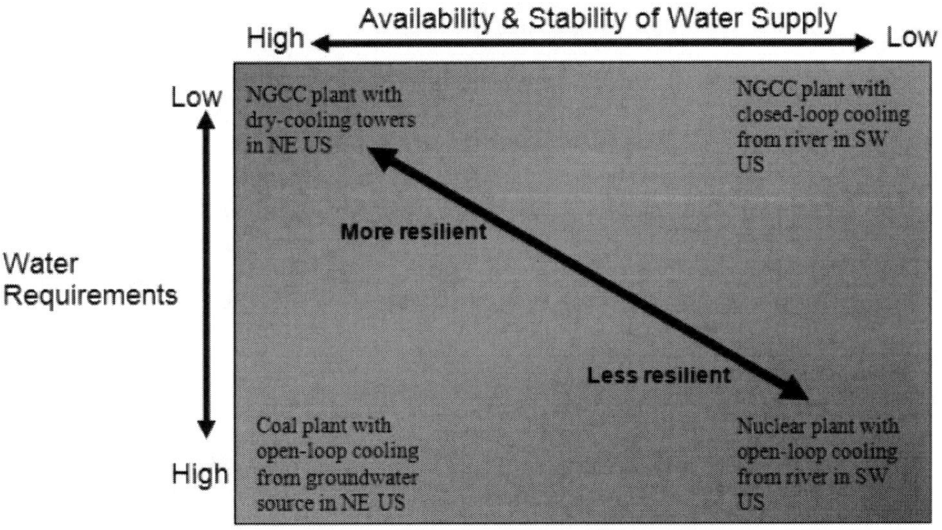

Figure 5-3. Example Results for Assessment of a Power Plant's Resilience to Droughts.

6. Mitigating Policy Options

Preparation for periods of drought is an important problem for water service agencies, households, and industry. Even under normal weather conditions, rapidly growing populations, environmental concerns, and expanding industries are increasing pressure on local water supplies. State and local governments are searching for policies to deal with the risks linked with uncertainty in water supplies and increasing demands.

6.1. Price and Non-price Restrictions on Water Use

Demand-side management (DSM) policies can be used to manage existing water supplies during drought. These non-price policies do not affect the residential price of water, but place direct controls on water use, such as rationing through varied means. Rationing, instituted as a strict policy where households are entitled to a fixed amount of water and then cut off from service, would likely prove unacceptable because water is considered a basic necessity. Non-price policies that decrease household demand for water are include water restrictions on selected uses, education campaigns to encourage voluntary conservation, and subsidies for adoption of water-efficient technologies. Water agencies could ban selected uses of water by prohibiting landscape irrigation or restricting watering of landscaping to selected days or times of day. Education campaigns could encourage particular types of watering techniques such as the use of hand irrigation or drip systems. Education could also inform households of the ideal time of day for outdoor water use as well as encourage overall changes to water-use behavior. Subsidy programs could offer rebates for the installation of low-flow toilets and low-flow showerheads, and, in the desert southwest, switching from swamp coolers to refrigerated air conditioners, using high Center for Energy and Environmental Resources (CEER)-rated compressors.

Price policies could be effective in curbing household water use by providing a monetary penalty/incentive for amounts of water use. Water agencies could implement a price policy that would move households from fixed, per unit, uniform rates to a moderately increasing block-price schedule. Another option would be to give each household a specific water allocation, with penalties for noncompliance. Price policies are considered regressive, because lower income households have a larger portion of their income allocated for water consumption than higher income households. Higher income households could also choose to simply pay penalties for non-compliance.

Demand-side water management non-price and price restrictions were implemented in California during the statewide drought from 1985 to 1992. Many policy instruments were used during this time, allowing for observation of consumer behavior to various policy instruments. Analyses of consumer behavior have been conducted for the communities of Santa Barbara and Goleta, selected for their differing policy regimes and socio-economic characteristics (MacDonnell 1990). Both communities used increasingly strict price and non-price policies to reduce demand. Household survey data, utility water use data, and price data were collected for these communities.

In 1986, Goleta promoted rebates for substituting new low-flow toilets for traditional toilets in homes and businesses; in 1987, the program was expanded to include low-flow showerheads. In 1988, Santa Barbara introduced a program dispensing free low-flow showerheads and rebates for replacing traditional toilets for low-flow toilets. In 1989, as the drought progressed, DSM price restrictions were introduced in both Santa Barbara and Goleta. Santa Barbara moved away from fixed per-unit uniform water rates to a moderately increasing block-price schedule. The pricing strategy for Goleta was based upon historical usage patterns; each household was entitled to a base water allocation per year, with marginal price penalties for households exceeding their allotment. In 1990, as water shortages grew more severe, more aggressive policies were needed to curb water demand. Santa Barbara prohibited some water uses such as all landscape irrigation and drip systems; a second price strategy of steeply increasing price blocks replaced the prior pricing strategy. Goleta responded to increased water scarcity by returning to uniform pricing rates, but at higher rates; specific water users were not penalized for their water usage.

During the 1985 – 1990 drought and DSM period, water demand trended downward. In Goleta, lower income households reduced their water consumption the most, while in Santa Barbara, higher income households reduced their consumption the most. For both Goleta and Santa Barbara, water demand declined for households with large yards. In the short run, it was found that a 10-percent increase in water price reduced aggregate demand by 3.3 percent. Because of the price changes, average household demand over the drought period fell 9.3 percent in Santa Barbara and 26.2 percent In Goleta. In the long run, for both Santa Barbara and Goleta, a 10- percent increase in prices resulted in a 3.9-percent reduction in demand (Archibald and Renwick, 1998b).

Across both locations, households with lower incomes responded more to higher water prices than wealthier household groups. For low income households, a 10-percent increase in water prices resulted in a 5.3-percent reduction in demand for water. Moderate to high income households decreased their demand by 2.2 percent. Wealthy households decreased their demand by the smallest amount, 1.1 percent.

In both Goleta and Santa Barbara rebate programs promoting low-flow showerheads and toilets decreased household water demand by 8 percent. Changes in irrigation technology reduced household water demand by 11 percent (Anderson 2008).

Non-price and price restrictions on water use for residential consumers will not have a cost impact for industrial users of water supplies.

6.2. Improve Agricultural Efficiency through Changes in Agricultural Techniques and New Technology

Drought and shrinking groundwater and surface water supplies are pressing consumers of water to make water supplies go further. As much as 70 percent of water used by farmers never makes it to the crops (*The Economist* 2008). The amount of water applied to crops can be adjusted, irrigation efficiency can be increased by better irrigation management, and land can be left idle. Farms with abundant and inexpensive water supplies often substitute water use for labor, management, or irrigation system investment.[9] Farms can switch from being completely reliant on groundwater or surface water sources to reliance on recycled water; recycling facilities can be built onsite and are typically complete within 2 years, which satisfies the 3-year-or-less criteria for our definition of short-term policy options.

Water is often lost by leaking irrigation channels, draining into rivers, or seeping back into groundwater. Improved irrigation efficiency would result in increased groundwater and surface water supplies for all users. Irrigation efficiency is a measure of the effectiveness of an irrigation system in delivering water to a crop and the effectiveness of irrigation in increasing crop yields. Water losses can be easily averted by switching to drip irrigation or repairing irrigation systems. All irrigation is subject to some amount of water loss; the intent of improved agricultural efficiency is to promote low-cost mitigating technologies that benefit agricultural, urban, and industrial users of water supplies.

Interruptible water markets would allow some water to move from irrigation to hydropower use in critical flow periods. Typically, irrigation use has priority regardless of how little remains for other uses when stream/river flow is low. Irrigation could still use the water in most years, but in dry periods, irrigation users would give up some water to maintain stream/river flows and power supplies. Hamilton and others believe interruptible water markets would require long-term contractual commitments, 20 to 25 years, to produce the most hydropower benefits (Hamilton, et al. 1989). Hamilton and others explain that these are long-term contracts because water supplies can range from normal to only half of normal consumptive use; for farmers, market participation would be an added source of uncertainty.

Recycled water is used in agriculture in many different parts of the world. The levels of treatment vary from almost no treatment to water that meets standards for drinking water. Recycled water comes from the urban sewer system and it goes through several purification treatments. The number and types of treatments the water receives determine the final quality of the water. Typically, water is put through two processes (secondary treatments) or three processes (tertiary treatments) and is then disinfected with chlorine, ultraviolet light, or reverse osmosis. California law (CCR Title 22, Chapter 3) currently allows tertiary-treated water and disinfected water to be used on all edible food crops.

In 2006 Watsonville, California, began building the Watsonville Recycled Water Facility (RWF) for agricultural use. The Watsonville RWF project is a joint project of the city of Watsonville and the Pajaro Valley Water Management Agency to reduce current levels of groundwater pumping and associated problems of seawater intrusion, while maintaining existing agricultural uses in the Pajaro Valley (Santa Cruz and Monterey counties). The Watsonville RWF was completed in December of 2008; it will provide 1.7 billion gallons of treated wastewater to local farmers every growing season (Anderson 2008). The planning for this project took approximately 8 years; with construction being complete in just 15 months. The combined cost of the RWF and its coastal distribution system was $65 million, with half paid through grants (Anderson 2008).

Improved agricultural efficiency is a relatively expedient way to lessen the impact of droughts. The associated costs of improved efficiency are the burdens of farmers or local municipalities. Those that benefit from improved agricultural efficiency are farmers, urban communities, and commercial/industrial users, particularly those reliant on water supplies for their processes. Commercial and industrial users incur no costs from improved agricultural efficiency; in fact, their water supply costs may decrease if improved agricultural efficiency results in increased groundwater and surface water supplies.

6.3. Use Reclaimed Water in Electric Power Generation as a Substitute for Traditional Sources of Water

The original source of reclaimed water is treated waste water that was once potable water. The significant difference being, total dissolved solids, which are typically higher in reclaimed water than in potable water. Differences in fresh water and reclaimed water qualities could make direct substitution for process cooling problematic. Potential users of reclaimed water would have to evaluate the possible effects of using reclaimed water on cooling system equipment protection and life expectancy because of cooling water-related corrosion, scaling, deposition, and biofouling (ANL 2007).

The City of Burbank Generating Station in Burbank, California, has six generating units. Cooling tower systems are used to cool the steam surface condensers. The City of Burbank Generating Station has used reclaimed water from the City of Burbank Waste Treatment Plant since 1966 (CCR Title 22, Chapter 3). Five of the six generating units use reclaimed water. Failures of copper tubing have reportedly been minimal; however, the use of reclaimed water at the City Burbank Generating Station has resulted in a chemical cost increase of approximately 20 percent (Selby 1996).

The Los Angeles Department of Water and Power's (DWP's) Scattergood Generating Station has three generating units. All three units use seawater for condenser cooling, but use an evaporating cooling tower system for service water. Scattergood makes use of seawater rather than potable water and is investigating incorporating the use of reclaimed water.

The costs for a cooling water treatment program, when supplied by a water treatment service company, depend upon a number of factors, including site-specific water equipment qualities, cooling system operation, and special client-vendor arrangements. A general cost estimate can be developed based on knowledge of the chemical costs, obtained from chemical suppliers, and chemical usage relative to cooling system water and equipment requirements. Reclaimed water contains phosphate and nitrate; these excessive phosphate levels could increase polymer costs by 20 to 50 percent (Selby 1996). Another factor is the cost of reclaimed water, which may be lower than potable water. In general, the entity responsible for generating electric power will also be responsible for the cost of using reclaimed water in the electric power generation process. The time required for switching technologies and the cost of implementing this technology may prove prohibitive.

6.4. Actions that Conserve Electricity and Water

Electric generation facilities that use water for cooling and/or for raising steam to produce electricity can conserve water by directly reducing the demand for or consumption of electricity.[10] In a short-term scarce water environment, electric utilities may need to examine all the options available to satisfy customer demand. There are several options that can be used to reduce electricity demand or consumption within the relevant 3-year timeframe. Many of these options can be implemented relatively rapidly and at relatively modest cost. Several of these options are discussed in the following sections:

While the savings for each individual residential, commercial, or industrial location might be modest, when added up across the community, service territory of a utility, or the whole country, these savings can add up to significant savings of electricity.

- **Managing the Demand Side:** During the 1980s and 1990s, the concept of DSM was widely discussed among electric utilities and energy experts, leading to implementation of DSM programs by electric utilities. The fundamental idea behind DSM is that reducing overall electricity demand and, in particular, the timing of that demand over the day or season, could be achieved at lower cost-per-unit output than the alternative strategy of increasing supply. This idea emanates from the significant peak in demand evidenced for electricity. Capacity is required to service peak demand during 4 to 8 hours each day and sits idle at other times. If the difference between peak and average demand could be reduced, not only could electric utilities be more profitable, but less capacity would be required.

  DSM programs include a wide array of possible actions that can be taken by end-use consumers of electricity. These actions, often induced by changes in utility policies and procedures and prices, include everything from window replacement and insulation programs to peak-load pricing of electricity to shift demand from peak consumption times to off-peak times. Some states forced the implementation of these programs as early as the mid-1970s in response to the first energy crisis (Eto 1996). While these programs fell out of favor towards the late 1990s and early 2000s, they may now be returning to favor, but in a different form. Regulated electric utilities were not particularly enthusiastic adherents to DSM programs, possibly because a reduction in consumption reduced revenues and the need for new plant capacity. In a rate-of-return regulatory environment, the more capital that is in the rate base, the more profits that are earned by the utility.

- **Extending and/or Reauthorizing Federal Tax Credits for Installation of Alternative Energy Technologies:** Solar power and wind power are rapidly coming closer to being cost-competitive with traditional coal-fired, natural gas-fired, and nuclear power generation. In fact, if full-cost accounting for carbon emissions were to be added to traditional electricity generation technology costs, it is likely that some alternative energy technologies would already be economically competitive. Nevertheless, given that there is no carbon-emissions legislation for traditional fossil fuels, tax credits could be used to accelerate implementation of alternative energy technologies. Already, many residential and commercial customers are installing photovoltaic solar collectors on businesses and homes (Rosenbloom 2008). Wal-Mart, Kohl's, Safeway, and Whole Foods Market are among the retailers already installing such systems. This trend was accelerating during late 2008, even considering the (then) uncertainty surrounding the continuation of incentive programs.[11] Some commentators have indicated that the possibility of loss of incentives is what is motivating the acceleration. However, the rate of implementation is indicative of the potential to make significant contributions to electricity supply with no increase in water consumption is readily apparent.[12]
- **Promoting and Creating Incentives for Electricity Conservation:** Electric utility companies may already have incentive programs to encourage their customers to weatherize and winterize their homes and businesses. Expanding and extending programs for window replacement, insulation, and installation of shades and shutters could pay dividends by direct reductions of electricity and natural gas consumption. These programs have the potential to reduce electric consumption within the timeframe of interest. Weatherizing and winterizing can save energy in both summer and winter. In summer, less leakage of cool air through window and door openings will save cooling costs. Shading windows from summer sun, by using special sun screens on the outside of windows, awnings, or overhangs, helps to conserve cooling costs. In winter, improved windows and doors save mostly natural gas and oil for home heating, but can also save electricity as fans for furnaces that distribute warm air run less.
- **Promoting conversion of incandescent to fluorescent light bulbs:** Incandescent light bulbs use only about 20 percent of the electricity they consume to produce light. The remaining electricity becomes heat, which then adds to space cooling load. In winter, the excess heat may actually displace some natural gas or other space-heating fuel consumption. Accordingly, significant electricity can be conserved by making the substitution. Incentives and promotions can be developed to encourage people to make the replacements.
- **Promoting lower thermostat settings in winter and higher settings in summer:** Natural gas is a main fuel for home heating in winter. Dropping the thermostat a few degrees can save significant energy. In summer, electricity drives the compressors that provide air conditioning. Increasing thermostat settings can significantly reduce peak electricity consumption during summer afternoons and early evenings. Shaving

the peak demand reduces the marginal capacity requirement for a summer peaking system.

For example, each lower degree setting of a thermostat in the winter and higher degree setting in the summer is approximately equivalent to 1 percent of a typical energy bill (Bond 1999). The following steps result in an estimate of the potential electric energy saved:

1. The Energy Information Administration (EIA) estimates that U.S. national average residential monthly energy consumption is 920 kilowatt-hours for 122.5 million customers, for a total national monthly consumption of 112.6 trillion kilowatt-hours of electricity (EIA 2009).[13]
2. One percent of the national monthly consumption calculated in step 1 is the potential savings from reducing thermostats by 1 degree, or 11.3 billion kilowatt-hours of electric consumption.
3. Dividing 11.3 billion kilowatt-hours from step 2 by the number of hours in a month (720) yields 156.5 million kilowatts; this is the number of kilowatts of output per month that could be saved by reducing all residential thermostats 1 degree.
4. This is equivalent to 156.5 thousand megawatts, which is the customary metric in which electric plant capacities are frequently quoted. At an assumed 75-percent average duty factor for each electric plant, over 200 electric plants (each with 1,000 megawatts of capacity) would be required to produce this instantaneous quantity of output.

Seemingly minor adjustments to consumption can have enormous effects when aggregated over millions of consumers. Additional savings in the commercial and industrial consumer sectors would add significantly to this total.

6.5. A Changing Future for Investor-owned Utilities

The decade of the 1990s was an exceptional period of consistently steady economic growth for the U.S. economy and monetary management that kept a fairly stable price level during this period of growth. This era has been identified as one of the truly remarkable periods in U.S. economic history.[14] World crude petroleum prices remained steady at reasonably comfortable low levels and energy prices generally were moderate. However, in the latter half of the decade, two issues arose that were to have a significant impact on investor-owned utilities (IOUs): "stranded costs" and the emerging recognition of global climate change and its root causes in the burning of fossil fuels.

The stranded-cost issue arose from the new technology of natural gas-fired combustion turbines that, at the time compared to coal and nuclear plants, were much easier to permit and faster to construct and, due to low natural gas prices, relatively low in fuel cost as well. These electric generation facilities were being planned by independent power producers (IPPs) whose existence had been contemplated by changes to energy policy regulations in the early part of the decade. In fact, IPPs were allowed to form precisely because the industry and energy policy makers expressed serious concern about the ability of the regulated utilities to increase generation capacity in response to increasing demand. IPPs are backed by private capital and exist outside the state electric utility regulatory environment. Nevertheless, regulated electric utilities were required, under the law, to purchase power from these facilities to avoid the cost of building coal and nuclear plants. The regulated utilities were concerned that these lower cost plants would "strand" their own coal and nuclear plants, which would become too expensive to operate. Therefore, utilities would be unable to pay off the debt incurred to build these plants. There is a direct correlation between the debate about stranded costs and the political forces demanding deregulation of the electric utility industry, including "unbundling" of generation, transmission, and distribution facilities to permit competition in parts of the industry. The *Public Utility Regulatory Policies Act* (PURPA) (16 U.S.C. §§ 2601 – 2645) was an additional factor causing IOUs concern for the financial soundness of their businesses. PURPA required alternative energy technologies, such as wind and solar power, to compete as supply sources based on the concept of "avoided cost." Avoided cost is defined as the marginal total cost of the next power plant facility in the utility's stack of available generation technologies.

More or less concurrently, the debate among scientists regarding global climate change emerged from the scientific literature and made its way into the public media. The causes of climate change are ascribed largely to carbon emissions caused by burning of fossil fuels. Pressure built to require electric utilities to take action to reduce carbon emissions with wide discussion of the possibility of a carbon tax. The difficulty, and therefore cost, of building large coal plants increased accordingly. Alongside difficulties in building commercial nuclear power plants due to concerns about their safety, utilities were left with relatively few viable options to increase plant capacity to meet increasing demand. The 1990s closed with a number of states undertaking serious efforts to deregulate their electric power industries and to establish markets to permit competition in the generation sector and in the retailing of electricity. Problems arose with many of these deregulation plans, and today, there is a confusing mix of traditional rate-of-return regulation and the use of markets, particularly at the wholesale level, in the supply and distribution of electricity.

Early in the 2000s, the increased use of natural gas-fueled combustion turbines began to increase the demand for natural gas. In response, the price of natural gas rose. Historically, the bulk of natural gas consumption was for space heating and the fuel consumption was seasonal, peaking in the winter. More recently, a larger portion of consumption is attributable to electric power generation, leading to a more level yearly consumption. This has caused prices to increase and changed the relative economics of combustion turbines, coal-fired, and nuclear plants.

The concern about global climate change, sustainability, and improvement in the engineering economics of alternative energy sources, primarily wind and solar, is combining with several of these other trends to create a sort of perfect storm for regulated electric utilities. Because it is difficult to get new coal and nuclear plants permitted, IPPs are developing large wind energy projects and contracting with utilities for the power these facilities will produce. Several new solar plants developed under the IPP business structure are under construction or in the planning stages.

Meanwhile, the concept of "net metering," which was required as a feature of the electricity deregulation provisions, has provided an opening for end-use customers to install photovoltaic cells and produce their own electricity, selling any excess into the grid. Thus, traditional electric utilities are squeezed from both ends and are becoming more like middle men who provide largely the transmission and distribution network to connect the IPPs that produce the supply and the end users. However, as more end users install photovoltaic cells, growth in demand for grid power is likely to decline.

6.6. Summary

It may appear that these policy options, which encourage conservation and energy efficiency, do not serve the interests of rate-of-return regulated electric utilities. After all, such business entities make their profits from the sale of electricity, and, the more electricity they sell, the greater their profits. While this may appear to be the case in the short term, in the long term, improving energy efficiency probably works to the advantage of the utility companies. Several factors lead to this possibility:

- Increased volume of profits in dollar terms might not necessarily mean an increased *rate* of profit. Profit rate increases depend both on the growth curve of total cost as well as of total revenue. In the short term, especially if there is excess capacity, it may be possible to increase the rate of profits by expanding output at little or no increase in average cost. Thus, the spread between average total revenue and average total cost might be maintained and the profit rate maintained. Eventually, however, demand and consumption will grow, requiring the installation of new capacity. This will increase marginal cost as well as average cost and reduce the profit rate. This is precisely the factor that makes conservation and energy efficiency the most economical strategy for electric utilities, as well as for the economy as a whole. So long as new capacity is higher in marginal cost than the average cost of the plants currently on the system, delaying the installation of new capacity will be profitable for the firm and beneficial for the economy.
- Energy inefficiency is likely to never be a viable or sustainable long-term strategy for a firm or an economy. To find verification for this requires looking no further than the current status of the U.S. auto industry. For years, U.S. automakers persisted in producing large, heavy, fuel-inefficient vehicles that, nevertheless, appeared to be profitable in the short term. Meanwhile, many foreign manufacturers pursued a strategy of smaller and more fuel-efficient autos and were inexorably stealing market share in the domestic U.S. market. When gasoline prices spiked, the resulting dramatic decline in demand for the large vehicles brought U.S. auto companies to the brink of bankruptcy. Thus, what appeared to be a profitable short-term strategy was unsustainable when considered in light of the long term. Because we can expect energy prices to increase on a trend basis, it is unlikely that energy inefficiency would ever be a viable long-term strategy.
- Beyond the possibilities outlined in the previous two bullets, an emerging recognition by utility regulators of the need to provide incentives to utilities to promote electricity conservation is leading to the implementation of mechanisms to decouple utility sales (from selling electric power) from earnings (McCarthy 2005). A number of states, including Connecticut, California, Oregon, Maine, and New York, have modified their regulatory statutes to encourage electric utilities to promote conservation and renewable energy technologies by providing financial rewards when a utility increases consumer education and promotes investments in

conservation programs and distributed generation. In California, an electric rate adjustment mechanism was developed to adjust actual revenues to a (presumably higher) amount that would reflect the effect of electricity *not* consumed as a result of conservation programs. The mechanisms vary, but an increasing number of states recognize that implementation of conservation and renewable results will be improved by aligning the utility's interests with the interests of the wider jurisdiction.

6.6.1. A New Business Model for Electric Utilities

The business model for a successful and profitable electric utility industry has been evolving for at least the last decade and probably longer. The evolution of this model received a significant boost from the passage of PURPA in 1997. This opened the market for independent power producers to develop electric generation facilities using private capital on a for-profit, non-regulated basis, and then enter contracts to sell the electricity to electric utilities. Today, many additions to generating capacity are developed by private capital, even those employing traditional steam-based technologies, using coal. Whether new nuclear generation capacity would be developed in this way isn't clear, due primarily to the significant technology risk associated with the open nature of the nuclear fuel cycle. Generally, private capital would have a preference for technologies that have less risk associated with them, particularly when there are a number of viable alternatives.

Most of the alternative energy generation sites developed in recent years and currently under development use this business model, involving IPPs who then sell the electricity to IOUs for transmission and distribution. Firms adept at planning, developing, and managing wind farms, for example, seek markets for the electric power output and sign contracts with utilities to purchase some minimum level of output from the planned facility. They then use these contracts to obtain financing for the project. The output of the facility, over and above the contracted amount, can then be sold on a merchant basis.

A new arrangement and opportunity for installation of solar photovoltaic cells is currently emerging across the country. Some progressive electric utilities are, in effect, renting the rooftops of residential, commercial, and industrial buildings (Galbraith 2008). Frequently, an unregulated subsidiary of an electric utility holding company will engage these contracts, paying the owners a periodic rental for the space. A variation on this theme could have the subsidiary front the capital to purchase and install the cells, thereby earning interest on the loan. Other aspects of the business opportunity could be to charge the building owner a fee to manage the cells. A key aspect of this large area of infrastructure development is that new transmission and distribution lines do not have to be constructed. Connectivity of producers and consumers of electricity is already an aspect of this system and the distribution of production and consumption locations around a wide area can promote system stability as well as a system security benefit.

It seems likely that many similar opportunities will emerge in the future as the nation makes a concerted effort to reformulate its energy system across the board. Most of the policy and technology options discussed in this paper are applicable to steam-powered plants, using water-based cooling technology located anywhere in the country. Sandia's focus has been predominantly on the western United States because that is where the drought conditions normally seem to exist. However, the drought maps contained in this report clearly indicate that droughts are possible east of the Mississippi River. While water-allocation mechanisms are somewhat different in the eastern United States, it is likely that mechanisms for transfer and reallocation are available and would simply need to be exercised.

REFERENCES

Anderson, Eric, 2008, "Water Recycling Project about 95 Percent Complete," *Pajaro Valley Register*, October 9

ANL (Argonne National Laboratory), 2007, "Use of Reclaimed Water for Power Plan Cooling," August

Archibald, Sandra O., and Mary Renwick, 1998a, "Expected Transaction Costs and Incentives for Water Market Development," in Easter, William K., et al., *Markets for Water: Potential and Performance*, Kluwer Academic Publishers, 95-117

Archibald, Sandra O. and Mary Renwick, 1998b, "Demand Side Management Policies for Residential Water Use: Who Bears the Conservation Burden?" *Land Economics*, 74(3): 343-359

Bond, Annie B., 1999, "10 Tips for the Thermostat: Your Key to Savings," *Care2,* August 21, 8:23 am, http://www.care2.com/greenliving/10-thermostat-tips-save-money.html

CCR Title 22, Chapter 3, Safe Drinking Water and Toxic Enforcement Act of 1986, *California Code of Regulation*

CEC (California Energy Commission), 2002, "Comparison of Alternate Cooling Technologies for California Power Plants: Economic, Environmental and Other Tradeoffs, Public Interest Energy," 500-02-079F, February

CEC (California Energy Commission), 2006, "Cost and Value of Water Use at Combined cycle Power Plants," CEC-500-2006-034, April

Cohen, G.E., D.W. Kerney, and G.J. Kolb, 1999, "Final Report on the Operation and Maintenance Improvement Program for Concentrating Solar Power Plants," SAND99-1290, Sandia National Laboratories, Albuquerque, New Mexico, June

Changnon, S.A., 1989, "The 1988 Drought, Barges, and Diversion," *Midwestern Climate Center Illinois State Water Survey*, Champaign, Illinois, 1989

Christie, Jim, 2008, "California water bank in works amid drought." Reuters Online Edition, September 4, 2008

DOE (U. S. Department of Energy), 2002, "Energy Penalty Analysis of Possible Cooling Water Intake Structure Requirements on Existing Coal-Fired Power Plants," National Energy Technology Laboratory, Argonne National Laboratory, October

DOE (U. S. Department of Energy), 2007, "Energy Demands on Water Resources: Report to Congress on the Interdependency of Energy and Water," U. S. Department of Energy, Washington D.C., January

EPRI (Electric Power Research Institute), 2002a, "U.S. Electricity Consumption for Water Supply and Treatment- the Next Half Century," *Water and Sustainability* (Volume 4), No. 1006787, Electric Power Research Institute, Palo Alto, California

EPRI (Electric Power Research Institute), 2002b, "U.S. Water Consumption for Power Production- the Next Half Century," *Water and Sustainability* (Volume 3), No. 1006786, Electric Power Research Institute, Palo Alto, California, March

EIA (Energy Information Administration), 2006, "Annual Energy Outlook 2006: With Projections to 2030," Energy Information Administration, Washington, D.C., February

EIA, (Energy Information Administration), 2007, "Electric Power Annual with Data for 2006," Energy Information Administration, Washington, D.C., October

EIA (Energy Information Administration), 2009, "Table 5. U.S. Average Monthly Bill by Sector, Census Division, and State 2007," report released January 2009, http://www.eia.doe.gov/cneaf/electricity/esr/table5.html

Eto, Joseph, 1996, "Utility Demand-Side Management Programs," Environmental Energy Technologies Division, Ernest Orlando Lawrence Berkeley National Laboratory, University of California, Berkeley, California, December

Explanation of the US Drought Monitor, http://drought.unl.edu/dm/classify.htm

Silviera, Ken, "Final RWF Project Statement," Ken Silviera, Manager, Wastewater Treatment Facilities, City of Watsonville, City Manager, Carlos Palacios

Galbraith, Kate, 2008, "Duke Energy to Rent Rooftops," Green, Inc. *Energy, the Environment and the Bottom Line*, December 11, http://greeninc.blogs.nytimes.com/2008/12/11/duke-energy-to-rent-rooftops/

Gardner, R.L., "The Potential for Water Markets in Idaho," *Idaho Economic Forecast* 7(1985):27-34

Hamilton, Joel R., Norman K. Whittlesey, and Philip Halverson, 1989, "Interruptible Water Markets in the Pacific Northwest," *American Journal of Agricultural Economics*, 71(1): 63-75

Hanak, Ellen, 2002. "California's Water Market, by the Numbers," Public Policy Institute of California, prepared for the Department of Water Resources, October

Hoffman, J., S. Forbes, and T. Feeley, 2004, "Estimating Freshwater Needs to Meet 2025 Electricity Generating Capacity Forecasts," National Energy Technology Laboratory, June

Howe, Charles W., 1998, "Spot Prices, Option Prices, and Water Markets: An Analysis of Emerging Markets in California," William K., et al. *Markets for Water: Potential and Performance*, Kluwer Academic Publishers, 11-140

Howitt, Richard, N. Moore, and R.T. Smith, 1992, "A Retrospective on California's 1991 Emergency Drought Water Bank," California Department of Water Resources, Sacramento, Caifornia

Howitt, Richard, 1998, "Expected Transaction Costs and Incentives for Water Market Development," In Easter, William K., et al. *Markets for Water: Potential and Performance*, Kluwer Academic Publishers, 95-117

Hutson, S.S., N.L. Barber, J.F. Kenny, K.S. Linsey, D.S. Lumia, and M.A. Maupin, 2004, "Estimated Use of Water in the United States in 2000," Circular 1268, United States Geological Survey, Washington, D.C., March

Leitner, A., 2002, "Fuel from the Sky: Solar Power's Potential for Western Energy Supply," NREL/SR-550-32160, National Renewable Energy Laboratory, Golden, Colorado, July

MacDonnell, Lawrence J., 1990, "The Water Transfer Process as a Management Option for Meeting Changing Water Demands," Natural Resources Law Center, School of Law, University of Colorado, Boulder, Colorado

McCarthy, Kevin E., 2005, "Decoupling Utility Sales and Earnings," OLR Research Report, 2005-R-0702, October 3, http://www.cga.ct.gov/2005/rpt/2005-R-0702.htm

Michelsen, Ari M., and Robert A. Young, 1993, "Optioning Agriculture Water Rights for Urban Water Supplies During Drought," *American Agriculture Economics Association*, 75: 1010-1020

NCDC (National Climactic Data Center), "Time Bias Corrected Divisional Temperature-Precipitation-Drought Index," http:://www1.ncdc.noaa.gov/pub/data/cirs/drought.README

NCDC Climate Monitoring Data, "Index of Pub/Data/Cirs," http://www1.ncdc.noaa.gov/pub/data/cirs/

Nieswiadomy, Michael L., and David J. Molina, 1989, "Comparing Residential Water Demand Estimates under Decreasing and Increasing Block Rates Using Household Data," *Land Economics* 65(3): 281-289.

Powicki, C.R., 2002, "The Water Imperative," in *EPRI Journal*, July.

Renwick, Mary, Richard Green, and Chester McCorkle, 1998, "Measuring the Price Responsiveness of Residential Water Demand in California's Urban Areas," Prepared for the California Department of Water Resources, May

Rosenbloom, Stephanie, 2008, "Giant Retailers Look to Sun for Energy Savings," NY Times Online Edition, August 11

"Running Dry," *The Economist*, September 18, 2008, website http://www.economist.com/world/international/displaystory.cfm?story_id=12260907

Schlumberger Limited, 2008, "Global Climate Change and Energy, Alternative Energy Sources: Hydroelectric Power," website http://www.seed.slb.com/en/scictr/watch/climate_change/alt_hydroelectric.htm, accessed May 19

Selby, Anthony K., Paul R. Puckorius, and Kris R. Helm, 1996, "The Use of Reclaimed Water in Electric Power Stations and Other Industrial Facilities," *Water, Air, and Soil Pollution*, 90: 183-193, July

SNL (Sandia National Laboratories), 2007, working draft of "Energy Water Challenges and Research and Development Priorities," Albuquerque, New Mexico, August 8

TVA (Tennessee Valley Authority), 2008a, "Hydroelectric Dam Diagram," website http://www.tva.gov/power/hydroart.htm, accessed May 19

TVA (Tennessee Valley Authority), 2008), "Pumped Storage Facility Diagram," website http://www.tva.gov/power/pumpstorart.htm, accessed May 19

U. S. Water News Online, 2006, "Low Water Levels Starting to Affect Mississippi River," website http://www.uswaternews.com/archives/arcsupply/6lowxwate9.html, accessed May 27

Visclosky, P.J., D.L. Hobson, H. Reid, and P. Domenici, 2004, Letter to Secretary of Energy Spencer Abraham, dated December 9, 2004, Washington, D.C., December

Wolken, L., 2006, "On-going Drought Threatens Missouri River Navigation," *Missouri Digital News* website http://www.mdn.org/2006/stories/barge2.htm, accessed May 27

33 U.S.C. (United States Code) §§ 1251-1387, *Federal Water Pollution Control Act (Clean Water Act*), October 18, 1972, as amended 1973-1983, 1987, 1988, 1990-1992, 1994, 1995 and 1996

16 U.S.C. (United States Code) §§ 2601 – 2645, *Public Utility Regulatory Policies Act*, Title 16, Chapter 46, 1978, as amended

APPENDIX A: DISCUSSION OF RECENT SOUTHEAST DROUGHT

From May 2007 through March 2008, much of the southeastern region of the United States saw an unusual level of drought, with Alabama seeing its driest year on record and the Atlanta area the second-driest on record.[15] Analysis of drought data from 1995 through 2008 shows that even in recent years, as the west saw very high levels of severe and extreme drought, very few regions in the southeast were affected. Table A-1 shows the climate regions in Alabama, Florida, Georgia, Mississippi, and Tennessee that saw severe and extreme drought at least 20 percent of the time at some point between 1995 and 2008. With the exception of the highlighted regions (North Central Florida and West Central Georgia), the area saw very little severe or extreme drought until the 2005 to 2008 period. The majority of the regions saw little to no instances of severe or extreme drought before that period.

Table A-1. Southeast climate regions with at least 20-percent severe or worse observations from 1995 to 2008 by percent of observation per time period

Name	State	1995-2008	1995-1999	2000-2004	2005-2008
Appalachian Mountain	AL	11%	0%	5%	35%
Northern Valley	AL	11	0	7	35
Eastern Valley	AL	11	0	5	35
Upper Plains	AL	9	0	5	28
Piedmont Plateau	AL	10	0	7	30
Gulf	AL	13	0	22	18
North	FL	15	5	12	35
Northwest	FL	11	2	5	33
North Central	FL	28	27	25	35
Northeast	GA	16	0	20	35
North Central	GA	6	0	0	23
Northwest	GA	11	0	5	35
East Central	GA	9	3	20	0
West Central	GA	30	18	35	40
South Central	GA	11	0	5	35
East Central	MS	9	0	8	23
Coastal	MS	13	0	20	20
Eastern	TN	7	0	0	28
Cumberland Plateau	TN	8	0	0	33
Middle	TN	8	0	5	23

Figure A-1 through Figure A-4 show the frequency of severe and extreme drought across the United States from 1995 to 2008 and broken up into five year increments, this data set is a subset of the data discussed in Section 4.1. Even the two climate regions that saw the most frequent occurrences of drought in the southeast are eclipsed by the frequency of drought in the west and barely visible in the data set.

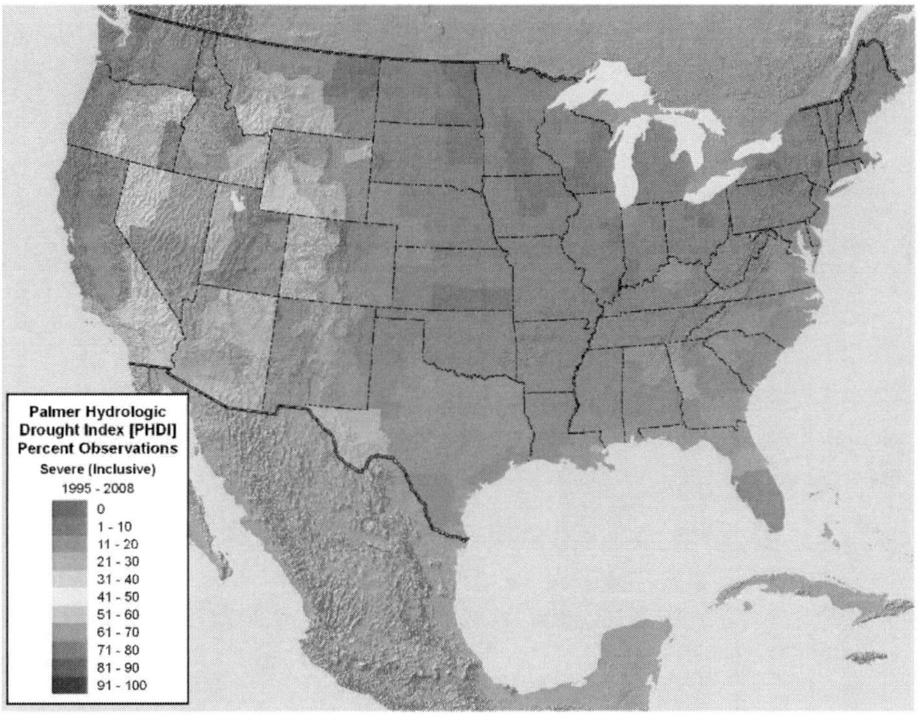

Figure A-1. Severe and Extreme Drought Frequency by Percent of Observations, 1995-2008.

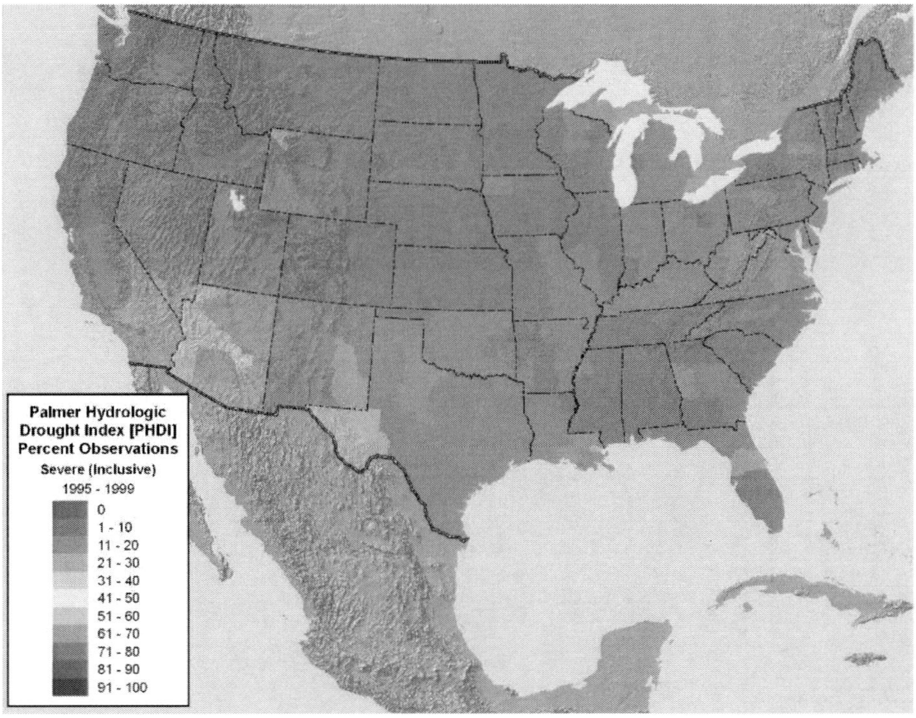

Figure A-2. Severe and Extreme Drought Frequency by Percent of Observations, 1995-1999.

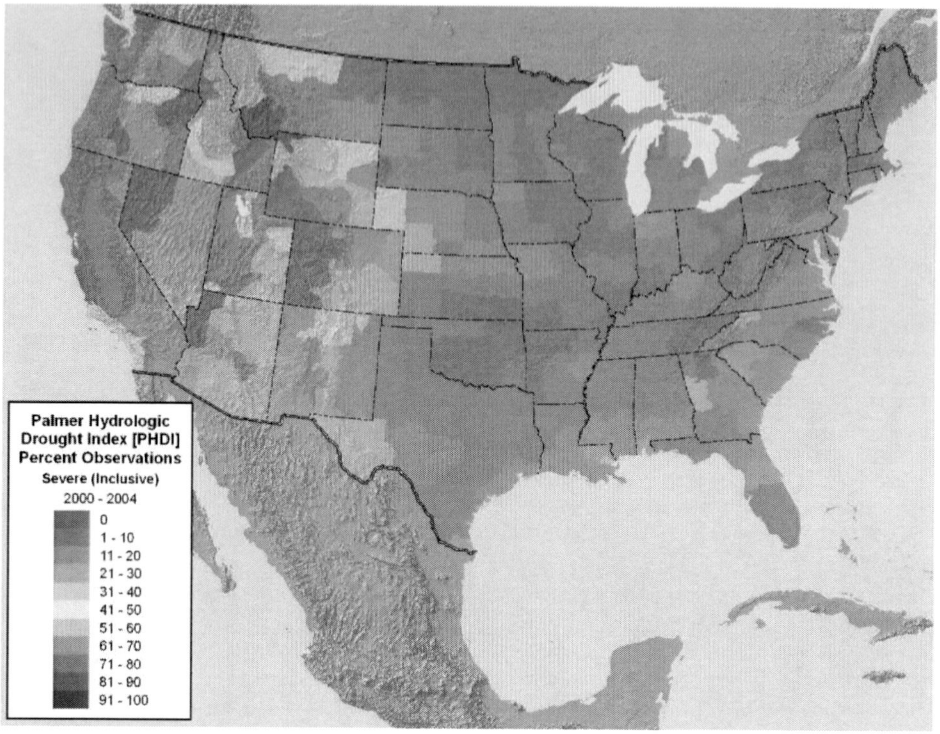

Figure A-3. Severe and Extreme Drought Frequency by Percent of Observations, 2000-2004.

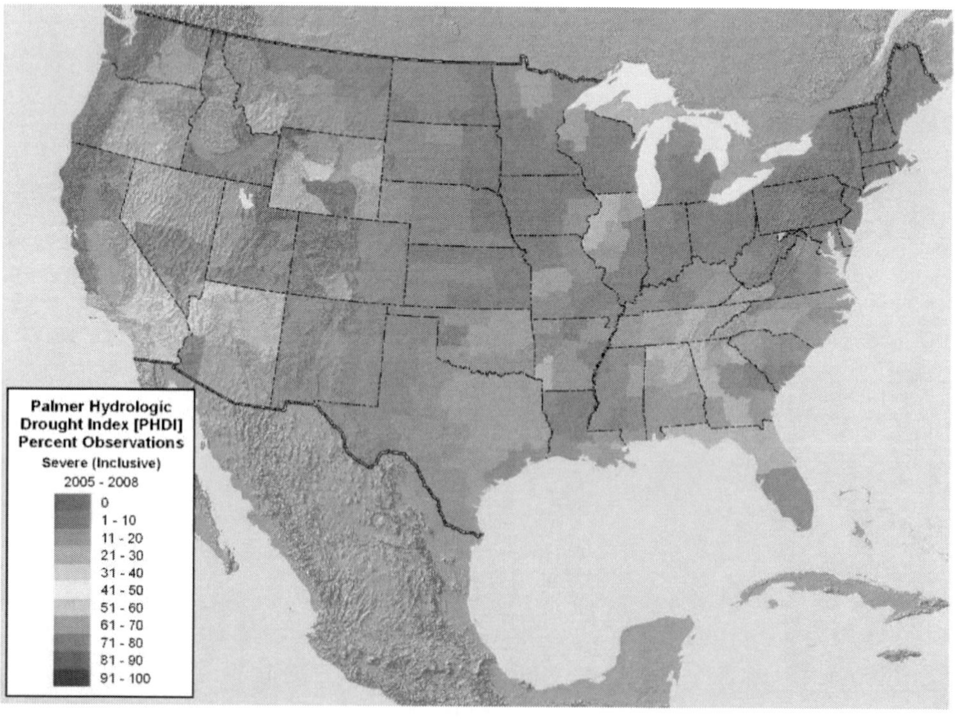

Figure A-4. Severe and Extreme Drought Frequency by Percent of Observations, 2005-2008.

End Notes

[1] In this document and DOE (2007), water consumption refers to the process of removing water from the hydrologic system in such a manner that it cannot be reused. Water "use" and "withdrawal" are terms that do not differentiate whether the water can be reused.

[2] Chapter IV of the congressional report provides several examples of these occurrences.

[3] While the quantity of water needed for plant operation is taken into account during plant-siting processes, the amount of water needed and the quantity of water available are not always correlated. This event may occur when other sources of water demand increase, thus decreasing the availability of water.

[4] National Climactic Data Center (NCDC), Time Bias Corrected Divisional Temperature-Precipitation-Drought Index, http:://www1.ncdc.noaa.gov/pub/data/cirs/drought.README

[5] NCDC Climate Monitoring Data, http://www1.ncdc.noaa.gov/pub/data/cirs/

[6] Explanation of the US Drought Monitor, http://drought.unl.edu/dm/classify.htm

[7] http://www.usatoday.com/weather/drought/2008-03-28-southeast-drought-eases_N.htm

[8] The primary technology that drives an electric generator, or converts energy to electricity

[9] When water is less expensive than labor and management, farming operations will choose not to employ laborers or farm management to monitor when crops and soil are in need of water and will, instead, water whether it is necessary or not.

[10] There is also an opportunity to change to technologies that don't use water in the production of electricity and that is one of the options suggested below. However, for some technologies, the planning and construction lead times are too long to fit the maximum three year criterion.

[11] Federal incentives have since been renewed.

[12] In September 2008, Congress authorized the extension of the existing tax credits.

[13]The average price was 10.4 cents per kWh in 2006 and average monthly expenditure was $95.66.

[14] Some analysts might argue that the seeds of the current financial troubles were sown in too liberal monetary policies and lack of regulatory oversight of the financial industry that began at the start of the 90s decade. Nevertheless, inflation during this period was moderate.

[15] http://www.usatoday.com/weather/drought/2008-03-28-southeast-drought-eases_N.htm

In: Dams, Drought and Energy-Water Interdependencies ISBN: 978-1-62808-942-4
Editor: Daniel L. Calzi © 2013 Nova Science Publishers, Inc.

Chapter 4

DROUGHT IN THE UNITED STATES: CAUSES AND ISSUES FOR CONGRESS*

Peter Folger, Betsy A. Cody and Nicole T. Carter

SUMMARY

Drought is a natural hazard with often significant societal, economic, and environmental consequences. Public policy issues related to drought range from how to identify and measure drought to how best to prepare for, mitigate, and respond to drought impacts, and who should bear associated costs. Severe drought in 2011 and 2012 fueled congressional interest in near-term issues, such as current (and recently expired) federal programs and their funding, and long-term issues, such as drought forecasting and various federal drought relief and mitigation actions. Continuing drought conditions throughout the country contribute to ongoing interest in federal drought policies and responses.

As of April 2013, drought has persisted across approximately two-thirds of the United States and is threatening agricultural production and other sectors. More than 1,180 counties so far have been designated as disaster areas for the 2013 crop season, including 286 counties contiguous to primary drought counties. In comparison, in August 2012, more than 1,400 counties in 33 states had been designated as disaster counties by the U.S. Secretary of Agriculture. Most attention in the 112th Congress focused on the extension of expired disaster assistance programs in separate versions of a 2012 farm bill. Attention in the 113th Congress again is expected to focus on farm bill legislation; however, other bills addressing different aspects of drought policy and response have also been introduced.

Although agricultural losses typically dominate drought impacts, federal drought activities are not limited to agriculture. For example, the 2012 drought raised congressional interest in whether and to what extent other federal agencies have and are using authorities to address drought. Similarly, the President in August 2012 convened the White House Rural Council to assess executive branch agencies' responses to the

* This is an edited, reformatted and augmented version of a Congressional Research Service publication, CRS Report for Congress RL34580, prepared for Members and Committees of Congress, from www.crs.gov, dated April 22, 2013.

ongoing drought. The Administration shortly thereafter announced several new administrative actions to address the drought.

While numerous federal programs address different aspects of drought, no comprehensive national drought policy exists. A 2000 National Drought Policy Commission noted the patchwork nature of drought programs, and that despite a major federal role in responding to drought, no single federal agency leads or coordinates drought programs—instead, the federal role is more of "crisis management." Congress may opt to revisit the commission's recommendations. Congress also may consider proposals to manage drought impacts, such as authorizing new assistance to develop or augment water supplies for localities, industries, and agriculture—or providing funding for such activities where authorities already exist. Congress also may address how the two major federal water management agencies, the U.S. Army Corps of Engineers and the Bureau of Reclamation, plan for and respond to drought.

This report describes the physical causes of drought, drought history in the United States, and policy challenges related to drought. It also provides examples of recurrent regional drought conditions.

INTRODUCTION

This report discusses how drought is defined (e.g., why drought in one region of the country is different from drought in another region), and why drought occurs in the United States. How droughts are classified, and what is meant by moderate, severe, and extreme drought classifications, are also discussed. The report briefly describes periods of drought in the country's past that equaled or exceeded drought conditions experienced during the 20th century. This is followed by a discussion of the future prospects for a climate in the western United States that might be drier than the average 20th-century climate. The report concludes with a primer on policy challenges for Congress, such as the existing federal/nonfederal split in drought response and management and the patchwork of drought programs subject to oversight by multiple congressional committees.

Following are brief answers to frequently asked questions related to drought.

What is drought? Drought is commonly defined as a lack of precipitation over an extended period, usually a season or more, relative to some long-term average condition. History suggests that severe and extended droughts are inevitable and part of natural climate cycles. While forecast technology and science have improved, regional predictions remain limited to a few months in advance.

What causes drought? The physical conditions causing drought in the United States are increasingly understood to be linked to sea surface temperatures (SSTs) in the tropical Pacific Ocean. Studies indicate that cooler-than-average SSTs have been connected to the severe western drought in the first decade of the 21st century, severe droughts of the late 19th century, and precolonial North American "megadroughts." The 2011 severe drought in Texas is thought to be linked to La Niña conditions (cooler-than-average SSTs) in the Pacific Ocean.

What is the future of drought in the United States? The prospect of extended droughts and more arid baseline conditions in parts of the United States could suggest new challenges to federal programs and water projects, which were conceived or constructed largely on the basis of 20th-century climate conditions. Some studies suggest a transitioning of the American West to a more arid climate, possibly resulting from the buildup of greenhouse gases in the atmosphere, raising concerns that the region may become more prone to extreme drought than it was in the 20th century. Some models of future climate conditions also predict greater fluctuations in wet and dry years; however, the net effect of such fluctuations is difficult to predict.

What is federal drought policy? Although drought impacts can be significant, no comprehensive national drought policy exists. Developing a national policy would be challenging because of split federal and nonfederal responsibilities; the existing patchwork of federal programs; and differences in regional conditions, risks, and available responses. In 2000, the National Drought Policy Commission provided recommendations to Congress to improve drought policy. Congress has acted on some of the recommendations (e.g., authorizing the National Integrated Drought Information System), but not others (e.g., creation of a National Drought Council and a fund to support drought planning). Given current conditions, Congress may review the functioning and adequacy of existing federal responses and programs (e.g., access to and level of assistance provided, incentives for mitigation of drought risk, and preparedness of federal facilities).

DROUGHT IN THE UNITED STATES—OVERVIEW

The likelihood of extended periods of severe drought, similar to conditions experienced centuries ago, and its effects on 21st-century society in the United States raise several issues for Congress. These issues include how to respond to recurrent drought incidents, how to prepare for future drought, and how to coordinate federal agency actions. For example, drought often results in agricultural losses, which can have local, regional, and national effects.

It also can affect other industries and services, including power and energy resource production, navigation, recreation, municipal water supplies, and natural resources such as fisheries, aquatic species, and water quality. How to address these impacts is an often recurring issue for Congress.

Addressing drought on an emergency basis is costly to individuals, communities, and businesses. Additionally, millions and sometimes billions of dollars in federal assistance can be expended in response to drought's social consequences. Thus, another recurrent policy issue is how to prepare and mitigate future drought impacts and how to do so efficiently across the many federal agencies with various and sometimes overlapping drought responsibilities.

Drought has afflicted portions of North America for thousands of years. Severe, long-lasting droughts may have been a factor in the disintegration of Pueblo society in the Southwest during the 13th century, and in the demise of central and lower Mississippi Valley societies in the 14th through 16th centuries.[1] In the 20th century, droughts in the 1930s (Dust Bowl era) and 1950s were particularly severe and widespread. In 1934, 65% of the contiguous United States was affected by severe to extreme drought,[2] resulting in widespread economic disruption and displacement of populations from the U.S. heartland.

Drought conditions are broadly grouped into five categories: (1) abnormally dry, (2) moderate, (3) severe, (4) extreme, and (5) exceptional.[3] Some part of the country is almost always experiencing drought at some level. Since 2000, no less than 6.6% of the land area of the United States has experienced drought of at least moderate intensity each year.[4] The land area affected by drought of at least moderate intensity varies by year and also within a particular year. For example, since 2000, the total U.S. land area affected by drought of at least moderate intensity has varied from as little as 6.6% (July 6, 2010) to as much as 55% (September 25, 2012). Based on weekly estimates of the areal extent of drought conditions since 2000, the average amount of land area across the United States affected by at least moderate-intensity drought has been 26%.

While the previous percentages refer to the extent of drought nationally, there is particular concern about those locations experiencing the most intense drought conditions. Nearly every year, *extreme drought*[5] affects some portion of the country. Since 2000, extreme drought or drier conditions have affected approximately 6.4% of the nation on average.[6] During August 2012, extreme drought extended over 20% of the country. Since 2000, exceptional drought conditions have affected approximately 1.4% of the nation on average. Of particular note were the conditions between June and October 2011; exceptional drought occurred over the largest land area—greater than 9%—during those months, with the affected areas concentrated in Texas.

This year, 2013, is likely to be another exceptional year in terms of the breadth of drought conditions throughout the country, particularly in the Great Plains and eastern portions of the Midwest.

The severe to exceptional drought conditions throughout the central and western parts of the United States appear to be persisting during spring 2013. (See Figure 1.)

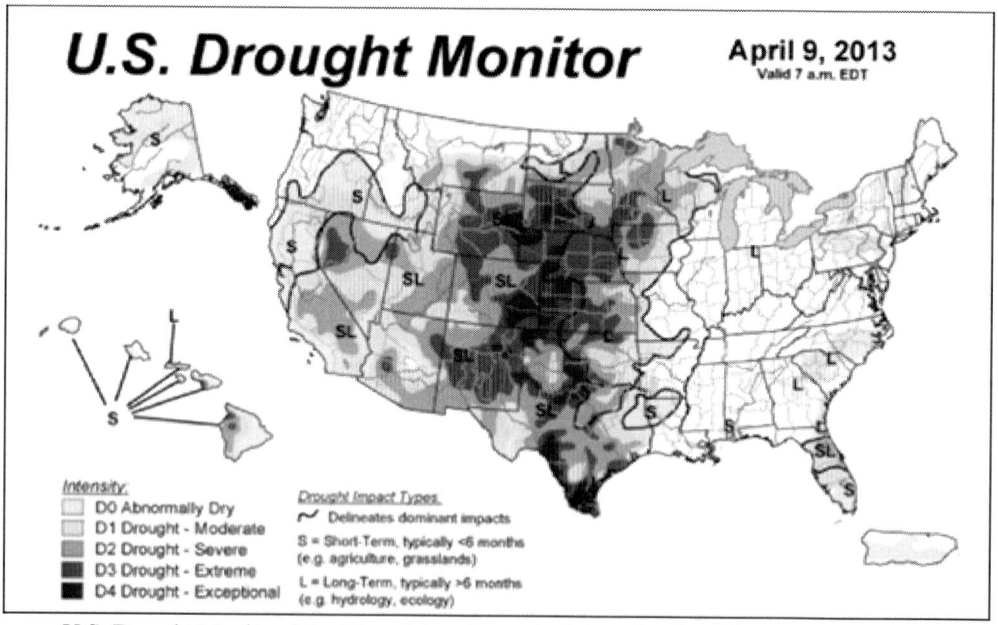

Source: U.S. Drought Monitor, http://droughtmonitor.unl.edu/monitor.html for April 9, 2013.

Notes: The areas delineated on the map as "drought impact types" depict regions where reports of specific impacts (e.g., short term (S) or long term (L) impacts) have been reported and tallied. For more information, see http://www.cpc.ncep.noaa.gov/products/predictions/tools/edb/droughtblends.php.

Figure 1. Extent of Drought in the United States.

DROUGHT IN NORTH AMERICA: SUMMER 2012 INTO 2013

In mid-August 2012, the extent of the drought conditions was significant: over 70% of the land area of the United States (including Alaska and Hawaii) was affected by abnormally dry and drought conditions.[7] The percentage of land area affected by abnormally dry or drought conditions stayed at or above 65% through February 2013, and remains above 60% as of mid-April, 2013.

The intensity of the drought varied across the country, but the regions of extreme and exceptional drought were clustered across the Midwest, Great Plains, Southwest, and in the Southeast, particularly Georgia in 2012. The widespread nature of drought and excessive heat conditions in the Midwest and Great Plains in the summer of 2012 contributed to sharply lower yields for major crops, including corn and soybeans. For example, the average U.S. corn yield was 123.4 bushels per acre, down more than 40 bushels from expectations before the onset of drought. For all crops, crop insurance indemnities for 2012 losses totaled $16 billion. In 2013, continuation of drought in the Great Plains and western parts of the Midwest is affecting prospects for wheat and other crops as well as pasture and forage conditions for livestock producers.

Figure 2 shows that Texas and portions of Florida and Georgia experienced exceptional drought conditions (the worst category of drought) in early 2012, while the upper Midwest, including most of the Mississippi Valley experienced normal conditions. A year later, in early 2013, the drought had eased somewhat in portions of Georgia and Florida, but intensified throughout the center of the country from Texas to the Canadian border. Nearly 12% of the contiguous United States was in exceptional drought conditions from late June 2011 through October 2011, compared to approximately 6% of the country the following year.[8] However, exceptional drought conditions persisted over nearly 6% of the contiguous United States from mid-August 2012 through mid-February 2013.

Although less severe for portions of the country, such as Texas and Florida, the 2012-2013 drought affected broader swaths of the agricultural heartland compared to 2011 (*Figure 2*). The 2012-2013 experience illustrates that the extent, timing, and particular features of areas affected by drought—dryland versus irrigated farm regions, or regions that are still recovering from previous droughts—are important in addition to the relative severity of drought conditions.

Origin of the 2012-2013 Drought

Figure 2 shows a snapshot of drought conditions for March in North America for 2010-2013. In 2010 most of North America and the United States were experiencing near-normal conditions. The extent and severity of the 2012-2013 drought raised questions regarding its origin, and whether the drought was consistent with the range of natural variability in the U.S. Midwest and Plains, or whether it was linked to longer-term changes in the Earth's climate system, such as human-induced global warming.

Although the images presented in Figure 2 may seem to indicate a steady progression of drought in the middle portion of the country from near-normal conditions in 2010 to widespread and intense drought in 2012-2013, a March 2013 analysis concludes that the 2012 intense drought was a discrete extreme event.[9] The report states that "the event did not appear to be just a progression or a continuation of the prior year's record drought event that developed in situ over the central U.S."[10] Instead, the report asserted that the drought developed suddenly, with near normal precipitation during winter and spring 2012 over the Great Plains.

The drought resulted from an extreme lack of precipitation during the summer months: 2012 was the driest summer in the historical record for the region, experiencing even less rainfall than the years 1934 and 1936, when the central Great Plains were about 0.5° C warmer than 2012.[11] Essentially the rains abruptly stopped in May over the central Great Plains, and did not return for the summer.

The report further stated that the 2012 summer drought was a "climate surprise," because summertime Great Plains rainfall has been trending upward since the early 20th century, and the last major drought occurred 25 years ago in 1988.[12] Further, the report concluded that neither sea surface temperatures, which have been rising generally due to global warming, nor changes in greenhouse gases in the atmosphere, were responsible for producing the anomalously dry conditions over the central Great Plains in 2012.[13] (A brief discussion of climate change and drought is below.)

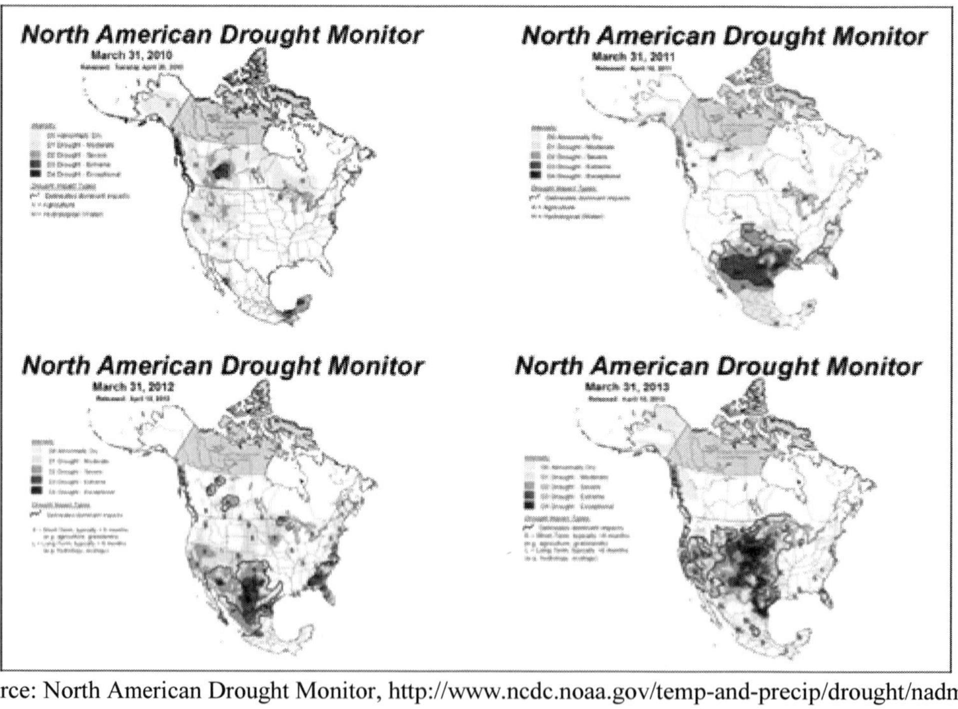

Source: North American Drought Monitor, http://www.ncdc.noaa.gov/temp-and-precip/drought/nadm/.

Notes: These drought intensity classifications represent broad-scale conditions. Local conditions may vary. The drought intensity classifications are discussed in detail in the text. Drought conditions are not analyzed in the gray shaded areas of northern Canada. Regions in northern Canada outside of the gray shaded area may not be as accurate as other regions in North America due to limited information.

Figure 2. Drought Conditions in North America, Comparing March 2010, 2011, 2012, and 2013.

WHAT IS DROUGHT?

Drought has a number of definitions; the simplest may be a deficiency of precipitation over an extended period of time, usually a season or more.[14] Drought is usually considered relative to some long-term average condition, or balance, between precipitation, evaporation, and transpiration by plants (evaporation and transpiration are typically combined into one term: evapotranspiration).[15]

An imbalance could result from a decrease in precipitation, an increase in evapotranspiration (from drier conditions, higher temperatures, higher winds), or both. It is important to distinguish between drought, which has a beginning and an end, and aridity, which is restricted to low rainfall regions and is a relatively permanent feature of an area's climate (e.g., deserts are regions of relatively permanent aridity).[16]

Higher demand for water for human activities and vegetation in areas of limited water supply increases the severity of drought. For example, drought during the growing season would likely be considered more severe—in terms of its impacts—than similar conditions when cropland lies fallow. For policy purposes, drought often becomes an issue when it results in a water supply deficiency: Less water is available than the average amount for irrigation, municipal and industrial supply (M&I), energy production, preservation of endangered species, and other needs.

These impacts can occur through multiple mechanisms: decreased rainfall and soil moisture affecting dryland farming; low reservoir levels decreasing allocations for multiple purposes (including irrigation, navigation, energy production, recreation, fish and wildlife needs, and other water supplies); low stream flows limiting withdrawals for multiple purposes, including municipal and industrial supplies, among others; decreased exchange of water in lakes resulting in water quality problems limiting recreation (e.g., blue-green algae restrictions in multiple lakes in Oklahoma and Texas during 2011 and 2012 drought conditions).

At the national level, drought is monitored and reported by the National Drought Mitigation Center in an index known as the U.S. Drought Monitor, which synthesizes various drought indices and impacts, and represents a consensus view of ongoing drought conditions between academic and federal scientists. Drought can also relate and contribute to other phenomena, such as fires and heat waves.[17]

Drought Is Relative

Drought and "normal" conditions can vary considerably from region to region. For example, the U.S. Drought Monitor shows that the southern tip of Texas, including Corpus Christi, faced extreme to exceptional drought in early April 2013 (*Figure 1*). Similarly, the city of Albuquerque, New Mexico, was in extreme drought during the same time period. However, Corpus Christi receives on average a total of 5.19 inches of precipitation over the three-month period January through March.[18] In contrast, Albuquerque receives on average 1.54 inches of precipitation over the same period.[19] Both cities faced extreme drought in early April 2013, but what was normal for Corpus Christi was very different for what was normal for Albuquerque.

To deal with these differences, meteorologists use the term meteorological drought—usually defined as the degree of dryness relative to some average amount of dryness and relative to the duration of the dry period. Meteorological drought is region-specific because atmospheric conditions creating precipitation deficiencies vary from region to region, as described above for Corpus Christi and Albuquerque.

Drought Is Multifaceted

In the past, U.S. Drought Monitor maps used an "A" to indicate that the primary physical effects are agricultural (crops, pastures, and grasslands) and an "H" to indicate that the primary impacts of drought are hydrological (to water supplies such as rivers, groundwater, and reservoirs). When both effects are apparent, the letters are combined, appearing as "AH." In the newer versions of the maps, such as the one shown in *Figure 1.* the "A" and "H" are replaced with an "S" and "L." These are experimental designations, according to the National Drought Mitigation Center, which produces the U.S. Drought Monitor maps.[20] The "S" designation is intended to indicate a combination of drought indices that reflect impacts that respond to precipitation over several days up to a few months (short-term effects). These would include impact to agriculture, topsoil moisture, unregulated streamflows, and aspects of wildfire danger.

The "L" designation approximates responses to precipitation over several months up to a few years (long-term effects). These would include reservoir levels, groundwater, and lake levels. *Figure 1* shows that for early April the nation, predominantly the western half, was experiencing a combination of all three types, S, L, and SL.

The U.S. Drought Monitor maps also indicate the intensity of a drought, ranging from abnormally dry (shown as D0 on the maps) to exceptional drought (shown as D4). How these conditions are assessed and how drought is classified is discussed below.

Drought Classification

To assess and classify the intensity and type of drought, certain measures, or drought indices, are typically used. Drought intensity, in turn, is the trigger for local, state, and federal responses that can lead to the flow of billions of dollars in relief to drought-stricken regions.[21] The classification of drought intensity, such as that shown in *Figure 1* for April 9, 2013, may depend on a single indicator or several indicators, often combined with expert opinion from the academic, public, and private sectors. The U.S. Drought Monitor uses five key indicators,[22] together with expert opinion, with indices to account for conditions in the West where snowpack is relatively important, and with other indices used mainly during the growing season.[23]

The U.S. Drought Monitor intensity scheme—D0 to D4—is used to depict broad-scale conditions but not necessarily drought circumstances at the local scale. For example, the large regions depicted as red in *Figure 1* faced extreme to exceptional drought conditions for the week of April 9, 2013, but they may contain local areas and individual communities that experienced less (or more) severe drought.[24]

Recent Examples: Texas, California, and Colorado River Basin

Drought in Texas, 2011-2013

In early April 2011, over 80% of Texas was experiencing severe to extreme drought, and nearly 10% of the state was in exceptional drought, the most severe level of drought intensity published by the National Drought Mitigation Center.[25] The 2011 drought in Texas represented a dramatic shift compared to the same time period in 2010, when approximately 4% of the total land area in Texas was experiencing drought conditions, with no exceptional drought conditions anywhere in the state. (See *Figure 3,* comparing 2010, 2011, 2012, and 2013.)

Drought and Electricity: Texas and the Pacific Northwest at Particular Risk

A December 2011 assessment of the drought vulnerability of electricity in the western United States revealed two regions whose electric generation was at risk—the Pacific Northwest and Texas.

The Pacific Northwest was shown to be vulnerable because of its heavy reliance on hydroelectric power generation.

In the summer of 2011, high heat in Texas resulted in increased demand for electricity, and power plants were operated for extended periods at maximum capacity. The operator of the grid that covers 75% of the state and 23 million people put into effect its emergency action alert system, which at first recommended conservation by customers but eventually deemed customer conservation critical to avoid rotating outages (rolling blackouts). In the end, only one plant on the Texas grid had curtailed generation due to water constraints; others were nearing curtailment when the weather conditions improved. The Texas grid was vulnerable because of heavy dependence on thermoelectric generation that relied on surface water for cooling and the region's high drought climate hazard.

Source: C.B. Harto et al., Analysis of Drought Impacts on Electricity Production in the Western and Texas Interconnections of the United States, Argonne National Laboratory, Oak Ridge, TN, December 2011, http://energy.s

Drought conditions worsened in Texas through the beginning of October 2011, when 88% of the state experienced exceptional drought conditions (and only 3% of the state was not classified as extreme or exceptional drought).[26]

Drought conditions generally improved throughout the rest of 2011, but large portions of the state were still affected by extreme or exceptional drought until late winter and early spring of 2012, when the eastern portion of the state recovered to normal or abnormally dry conditions (the least severe category) because of above-normal rainfall from December 2011 through February 2012.[27]

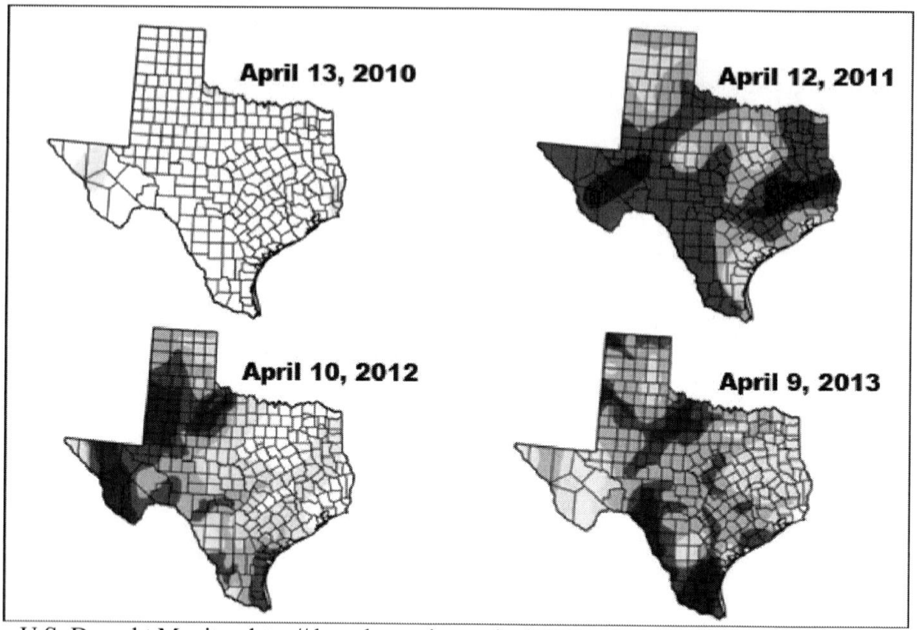

Source: U.S. Drought Monitor, http://droughtmonitor.unl.edu/, Modified by CRS.
Notes: See Figure 1 for explanation of colors shown on maps.

Figure 3. Comparison of Drought Conditions in Texas in 2010, 2011, 2012, and 2013.

In April 2012 the drought in Texas had eased somewhat; slightly more than 50% of the state was experiencing severe drought or worse, down from 80% a year before. Conditions in April 2013 have worsened compared to a year earlier, however; approximately 70% of the state was experiencing at least severe drought in April 2013, and extreme and exceptional drought were 30% and 12%, respectively.[28] (See *Figure 3.)*

According to Texas state climatologist John Nielsen-Gammon, 2011 may have been the worst one-year drought on record for Texas.[29] Compounding the effects of abnormally low precipitation, the June-August average temperature in Texas was approximately 2.5 degrees Fahrenheit greater than any previous Texas summer since 1895 and 5 degrees Fahrenheit (F) greater than the long-term average.30 The 2011 U.S. Drought Monitor showed that Texas had been experiencing both hydrological and agricultural drought, indicating that the drought had caused deficiencies in water supplies as well as deficiencies of water to crops, plants, and grasses.[31]

Drought Conditions Affecting the Rio Grande Project in 2011 and 2012

The 2011-2013 drought conditions in Texas and the Southwest have affected the amount of water in the Rio Grande river, which flows south from Colorado through New Mexico to form the U.S. border between Texas and northern Mexico. The U.S. Bureau of Reclamation's Rio Grande Project, which furnishes irrigation water for approximately 178,000 acres in New Mexico and Texas, as well as electric power, includes the Elephant Butte dam and reservoir and the Caballo dam and reservoir. Both dams and reservoirs are in New Mexico, and about 60% of the lands receiving irrigation water are in New Mexico.

Elephant Butte dam and reservoir provide year-round electric power generation and water during the irrigation season. Water released from Elephant Butte during winter power generation is stored downstream in the Caballo reservoir for irrigation use during the summer. About 40% of the lands receiving water from the project are in Texas, and water is also provided per a treaty with Mexico to irrigate about 25,000 acres in the Juarez Valley. The timing of the water releases in 2012 for delivery to Mexico and their potential impacts on U.S. regional interests (e.g., potential conveyance losses because releases for Mexico would not be timed with deliveries to U.S. water districts) raised concerns among some U.S. stakeholders about how scarce regional water resources are to be managed during dry conditions. Mexican growers sought the surface water deliveries because pumping problems had impaired their ability to start the agricultural season using groundwater. Inflow to Elephant Butte reservoir in 2011 was less than 15% of the 30-year average for March through July and is expected to be 39% for 2013; however, due to low runoff averages since 2008, reservoir storage is extremely low. For example, combined Elephant Butte and Caballo reservoir storage is 189,342 acre-feet—8.5% of combined capacity. The 2011-2013 drought conditions have exacerbated low flows into the reservoir; flows into the reservoir have exceeded average runoff values only three times in the past 15 years (1997, 2005, and 2008). Further, more than half the current available storage is "credit water" owned by upstream users and therefore not available for downstream use. Thus, only 4% of the current combined storage capacity will be available for downstream use. In 2011, water deliveries from the Rio Grande Project to the irrigation district in Texas (El Paso County Water Improvement District No. 1) as well as to the city of El Paso were stopped a month earlier than normal and the New Mexico portion of the project, operated by the Elephant Butte Irrigation District, stopped taking surface water deliveries in mid-July 2011. Most of these farmers turned to groundwater withdrawals to offset lost surface water supplies. Rio Grande Project water users were receiving a 20% allocation of water supplies as of April 1, 2012, and ended the year with a 38.7% allocation for the year. However, the preliminary allocation for 2013 is 4.2% as of March 31, 2013.

The dry conditions in the basin along with efforts to try to restore the populations of the endangered silvery minnow in some New Mexico reaches of the Rio Grande have focused attention not only on Bureau of Reclamation facilities and efforts in the basin, but also on whether a more comprehensive recovery implementation plan for the species might augment storage at other facilities to increase the flexibility of releases to the river. The Army Corps of Engineers also operates federal projects in the basin that currently are principally operated for flood control.

Sources: U.S. Bureau of Reclamation, Rio Grande Project, http://www.usbr.gov/ projects/ Project.jsp?proj_Name= Rio+Grande+Project; Texas Agrilife Research Center at El Paso, "Drought Watch on the Rio Grande," September 2, 2011, May 1, 2012, and February 8, 2013; email from Dionne Thompson, Chicf, Congressional and Legislative Affairs, U.S. Bureau of Reclamation, September 15, 2011; personal communication with Filiberto Cortez, Division Manager, El Paso Field Office, U.S. Bureau of Reclamation, May 2, 2012, and April 18, 2013; letter from Patrick R. Gordon, Texas Commissioner, Rio Grande Compact Commission, to Edward Drusina, Commissioner, United States Section, International Boundary and Water Commission, April 9, 2012; Texas Agrilife Research Center "Drought Watch, February 2013.

The most severe Texas drought overall occurred from 1950 to 1957, and had substantial impacts on water supplies across the state because it lasted over many years. Because of the longevity and severity of the 1950s drought, municipal water supplies in Texas today are designed to withstand a drought of similar magnitude, according to the state climatologist. Long-term precipitation patterns in Texas are influenced by a configuration of sea surface temperatures known as the Pacific Decadal Oscillation (PDO). Similar conditions also prevailed from the 1940s through the 1960s, encompassing the Texas drought of record (1950-1957).

The 2007-2009 California Drought and Outlook for 2013

The 2007-2009 California drought[32] was complicated by decades of tension over water supply deliveries for irrigation and M&I uses, and the preservation of water flows to protect threatened and endangered species. Dry conditions that began in 2007 continued through the 2009 water year (October 2008 through September 2009) and into the fall of 2009. According to the California Department of Water Resources, the 2007-2009 drought was the 12th-driest three-year period in California history since measurements began.[33] Although hydrological conditions were classified as below normal in 2010 and "wet" (well above average) in 2011, the 2012 water year was classified as "below normal" for the Sacramento River basin and "dry" for the San Joaquin River basin.[34] Above-average reservoir storage at the end of 2011 mitigated reductions to water users. Although the drought was declared over[35] in spring 2011, by August 2012, the U.S. drought monitor again showed increasing severity of drought in the eastern portion of the state. Water deliveries to state and federal water project contractors were restricted again in 2012.[36]

California's dry conditions from 2007 through 2009 exacerbated an already tight water supply, where federal and state water deliveries had been reduced in response to a court order to prevent extinction of the Delta smelt.[37] Similar factors are still in play today. Water deliveries from state and federal water projects for 2013 are restricted due to legal actions to protect threatened and endangered species, water quality requirements, and hydrological factors. Some federal agricultural water supply contractors are projected to receive just 20% of contract water supplies from the Bureau of Reclamation's Central Valley Project as of April 5, 2013, while some municipal and industrial contractors are scheduled to receive 70% of contract supplies.[38] The April 9, 2013, drought monitor shows abnormally dry to severe drought conditions in almost every corner of the state.

Conditions in the Colorado River Basin

Spanning parts of Arizona, California, Colorado, New Mexico, Nevada, Utah, and Wyoming, the Colorado River basin is a critical water supply for the West and portions of northwestern Mexico. Based on inflows observed over the last century, the river is over-allocated, and some contend that supply and demand imbalances are likely to increase in the future.[39] The March 2013 forecast is for flows below average for 2013, as has been the case for 10 of the last 13 years. The 2013 forecast for Lake Powell, the basin's largest reservoir, was for half of its normal inflow as the result of below average snowpack; that is, the inflow is anticipated at 5.3 million acre-feet. As called for in the water compact and interim guidelines that prescribe water sharing in the basin, the Bureau of Reclamation will most likely release from Lake Powell 8.2 million acre-feet from the reservoir for downstream purposes, thus resulting in a net drop in the lake level by the end of the water year. The dry conditions in 2013 are following the dry conditions of 2012. If forecasted conditions transpire, the basin will have storage levels in September 2013 near the historic lows of 2005. Reclamation predicts that Lake Powell storage levels will be 44% of capacity—10.78 million acre-feet—at that time.[40]

Drought in part of the basin, particularly the upper basin, which is the source of most of the river's flow, exacerbates tensions over the sharing of the resource and results in difficult tradeoffs among the multiple uses of water (e.g., municipal, agricultural, hydropower, energy, recreation, and ecosystem and species demands). How water resources are allocated among these uses within a state is largely determined by state water law, compliance with federal and state laws (including interstate compacts, and environmental and resource management laws and regulations), and court decisions. In the case of the Colorado River, apportionment of water supplies among the seven basin states is done in accordance with the Colorado River Compact and a body of law known as the "Law of the River." [41]

Low water availability in the Colorado River basin has effects beyond the basin boundaries. For example, Colorado River water is transported from Colorado's Western Slope to the state's Front Range; this water represents a significant contribution to the water available for agricultural and municipal uses in many eastern Colorado counties. Similarly, much of the lower basin allocation (4.4 million acre-feet) is diverted to southern California under the Colorado River Compact.

WHAT CAUSES DROUGHT IN THE UNITED STATES?

The immediate cause of drought is:

> the predominant sinking motion of air (subsidence) that results in compressional warming or high pressure, which inhibits cloud formation and results in lower relative humidity and less precipitation. Regions under the influence of semi permanent high pressure during all or a major portion of the year are usually deserts, such as the Sahara and Kalahari deserts of Africa and the Gobi Desert of Asia.[42]

Prolonged droughts occur when these atmospheric conditions persist for months or years over a certain region that typically does not experience such conditions for a prolonged period.[43]

Predicting drought is difficult because the ability to forecast surface temperature and precipitation depends on a number of key variables, such as air-sea interactions, topography, soil moisture, land surface processes, and other weather system dynamics.[44] Scientists seek to understand how all these variables interact and to further the ability to predict sustained and severe droughts beyond a season or two in advance, which is the limit of drought forecasting abilities today.

In the tropics, a major portion of the atmospheric variability over months or years seems to be associated with variations in sea surface temperatures (SSTs). Since the mid- to late

1990s, scientists have increasingly linked drought in the United States to SSTs in the tropical Pacific Ocean. Cooler than average SSTs in the eastern tropical Pacific region—"la Niña-like" conditions—have been shown to be correlated with persistently strong drought conditions over parts of the country, particularly the West.[45] A number of studies have made the connection between cooler SSTs in the eastern Pacific and the 1998-2004 western drought,[46] three widespread and persistent droughts of the late 19th century,[47] and past North American "megadroughts" that occurred between approximately 900 and 1300 A.D.[48] The precolonial megadroughts apparently lasted longer and were more extreme than any U.S. droughts since 1850, when instrumental records began. Some modeling studies suggest that within a few decades the western United States may again face higher base levels of dryness, or aridity, akin to the 900- 1300 A.D. period.[49] Although the relationship between cooler than normal eastern tropical Pacific SSTs (La Niña-like conditions) and drought is becoming more firmly established, meteorological drought is probably never the result of a single cause. Climate is inherently variable, and accurately predicting drought for one region in the United States for more than a few months or seasons in advance is not yet possible because so many factors influence regional drought. What is emerging from the scientific study of drought is an improved understanding of global linkages—called teleconnections by scientists—between interacting weather systems, such as the El NiñoSouthern Oscillation, or ENSO. (See box for a description of ENSO.) For example, some scientists link La Niña conditions between 1998 and 2002 with the occurrence of near-simultaneous drought in the southern United States, Southern Europe, and Southwest Asia.[50]

El Niño-Southern Oscillation (ENSO)

Under normal conditions, the trade winds blow toward the west in the tropical Pacific Ocean, piling up the warm surface waters so that the ocean surface off Indonesia is one-half meter higher than the ocean off Ecuador. As a result, deep and cold water flows up to the surface (upwelling) off the west coast of South America. The upwelling waters are 8 degrees Celsius (14.4 degrees Fahrenheit) cooler than waters in the western Pacific. During El Niño, the trade winds relax, upwelling off South America weakens, and sea surface temperatures rise. The El Niño events occur irregularly at intervals of 2-7 years, and typically last 12-18 months. These events often occur with changes in the Southern Oscillation, a see-saw of atmospheric pressure measured at sea level between the western Pacific and Indian Ocean, and the eastern Pacific. Under normal conditions, atmospheric pressure at sea level is high in the eastern Pacific, and low in the western Pacific and Indian Oceans. As implied by its name, the atmospheric pressure oscillates, or see-saws, between east and west; and during El Niño the atmospheric pressure builds up to abnormally high levels in the western tropical Pacific and Indian Oceans—the El Niño-Southern Oscillation, or ENSO. During a La Niña, the situation is reversed: Abnormally high pressure builds up over the eastern Pacific, the trade winds are abnormally strong, and cooler-than-normal sea surface temperatures occur off tropical South America. Scientists use the terms ENSO or ENSO cycle to include the full range of variability observed, including both El Niño and La Niña events.

Source: Tropical Ocean Atmosphere Project, Pacific Marine Environmental Laboratory, at http://www.pmel.noaa.gov/tao/proj_over/ensodefs.html.

Prehistorical and Historical Droughts in the United States

Some scientists refer to severe drought as "the greatest recurring natural disaster to strike North America."[51] That claim stems from a reconstruction of drought conditions that extends back over 1,000 years, based on observations, historical and instrumental records where available, and on tree-ring records or other proxies in the absence of direct measurements.[52] What these reconstructions illustrate is that the coterminous United States has experienced periods of severe and long-lasting drought in the western states and also in the more humid East and Mississippi Valley. The drought reconstructions from tree rings document that severe multidecadal drought occurred in the American Southwest during the 13th century, which anthropologists and archeologists suspect profoundly affected Pueblo society. Tree ring drought reconstructions also document severe drought during the 14th, 15th, and 16th centuries in the central and lower Mississippi Valley, possibly contributing to the disintegration of societies in that region.[53]

More recently, a combination of tree ring reconstructions and other proxy data, historical accounts, and some early instrumental records identify three periods of severe drought in the 19th century: 1856-1865 (the "Civil War drought"), 1870-1877, and 1890-1896.[54] The 1856-1865 drought, centered on the Great Plains and Southwest, was the most severe drought to strike the region over the last two centuries, according to one study.[55] The 1890-1896 drought coincided with a period in U.S. history of federal encouragement of large-scale efforts to irrigate the relatively arid western states under authority of the Carey Act.[56] Congressional debate also occurred over a much larger federal role in western states irrigation, which led to the Reclamation Act of 1902.

In the 20th century, the 1930s "Dust Bowl" drought and the 1950s Southwest drought are commonly cited as the two most severe multiyear droughts in the United States.[57] (The 1987-1989 drought was also widespread and severe, mainly affecting the Great Plains but also instigating extensive western forest fires, including the widespread Yellowstone fire of 1988.) According to several studies, however, the 19th and 20th century severe droughts occurred during a regime of relatively less arid conditions compared to the average aridity in the American West during the 900 to 1300 A.D. megadroughts. One study indicates that the drought record from 900 to 1300 A.D. shows similar variability—drought periods followed by wetter periods—compared to today, but the average climate conditions were much drier and led to more severe droughts.[58]

DROUGHT AND CLIMATE CHANGE

The relationship between climate change and future trends in droughts is complex and its scientific understanding appears to be evolving. In 2007 the Intergovernmental Panel on Climate Change (IPCC) released its Fourth Assessment Report, which stated that, globally, very dry areas have more than doubled since the 1970s due to a combination of El Nino-Southern Oscillation (ENSO)[59] events and global surface warming.[60] The IPCC report added that very wet areas declined by about 5% globally. The report asserted that documented trends in severe droughts and heavy rains show that hydrological conditions are becoming more intense in some regions.

In 2012, however, the IPCC issued a new report that noted: "There are still large uncertainties regarding observed global-scale trends in droughts."[61] The new report noted that its earlier assessment, that very dry areas have more than doubled since the 1970s, was based largely on only one study, which relied on a measurement largely related to temperature, not moisture. A different study, which looked at soil moisture simulations, found that trends in drought duration, intensity, and severity predominantly were *decreasing*, not increasing, but with strong regional variation.[62]

The 2012 IPCC report assigned medium confidence[63] that there has been an overall slight tendency toward *less* dryness in North America (i.e., a wetting trend with increasing soil moisture and runoff). It noted that the most severe droughts in the 20th century occurred in the 1930s and 1950s, where the 1930s drought was the most intense and the 1950s drought was the most persistent. In comparison to the severe megadroughts that occurred in North America hundreds and thousands of years ago, as documented using paleoclimate evidence (discussed elsewhere in this report), these recent droughts were not unprecedented, according to the 2012 IPCC report.

The report concluded that despite new studies that have furthered the understanding of mechanisms leading to drought, there is still limited evidence to attribute observed changes. The IPCC assessed that there was medium confidence that anthropogenic influence has contributed to changes in drought patterns in the second half of the 20th century, but gave low confidence to the attribution of changes in drought patterns at the regional level.[64] The report noted that some regions of the world have experienced trends towards more intense and longer droughts, such as southern Europe and West Africa. But in other regions, such as central North America and northwestern Australia, droughts have become less frequent, less intense, or shorter. How the 2011-2012 drought in the central United States may change that assessment in the forthcoming IPCC report (2014) remains to be seen.

Further adding to the complexity and challenge to the scientific understanding of what causes drought was work presented at the European Geosciences Union meeting in April 2013 that attempted to simulate megadroughts that occurred in the past.[65] The simulations produced a number of megadroughts that lasted for decades; however, they did not match the timing of the past documented megadroughts. The scientists presenting their work at the meeting concluded that the model they used seemed to miss some of the dynamics that drive large droughts.

RESPONDING TO AND PLANNING FOR DROUGHT

Several recent droughts triggered federal responses. When a drought is declared for a locality or region by the U.S. President, U.S. Secretary of Agriculture, or a state governor, it sets in motion a series of alerts, recommendations, activities, and possible restrictions at the local, regional, or state level, depending on the drought length and severity. Ultimately, a drought could initiate a federal response and transfer of federal dollars to the affected area.

Before drought severity reaches a level triggering a federal response, many states take action. The National Drought Mitigation Center posts online copies of drought management, mitigation, or response plans for states and localities, nationwide.[66] The California and Texas governors also have in recent years issued state drought emergency declarations triggering state drought assistance. Some states have also instituted water banks and water transfer mechanisms to deal with water supply shortages (e.g., California, Idaho, and Texas).

Federal Aid

When a state's resources are lacking, a state governor may request drought disaster assistance through the U.S. Secretary of Agriculture, who can declare an agricultural disaster due to drought and make available low-interest loans for qualified farmers and ranchers and other emergency assistance. However, under a new Farm Services Agency (FSA) streamlined process, any portion of a county experiencing severe drought according to the U.S. Drought Monitor for eight consecutive weeks can receive a "nearly" automatic USDA disaster declaration.[67] Further, any county for which a portion is identified in the U.S. Drought Monitor as undergoing severe drought (or worse) may also be declared a disaster area.

As of August 13, 2012, more than 1,400 U.S. counties in 33 states had been designated as drought disaster areas by the Secretary.[68] As of April 17, 2013, more than 1,180 counties had been designated as drought disaster areas for the 2013 crop loss purposes (primary and contiguous county designations).[69] For more on emergency assistance, see the box "USDA Emergency Assistance: Status and Legislation."

In addition, if the effects of a drought overwhelm state or local resources, the President, at the request of the state governor, is authorized under the Stafford Act (42 U.S.C. 5121 et seq.) to issue major disaster or emergency declarations resulting in federal aid to affected parties.[70] However, the last presidential drought or water shortage disaster declaration in the continental United States was for New Jersey in 1980. More recent drought declarations have been issued for U.S. territories in the Pacific. The infrequency of presidential domestic drought declarations increases the uncertainty about the circumstances under which such a declaration is likely to be made. The de facto policy since the 1980s has been that the U.S. Secretary of Agriculture is the lead in responding and declaring drought and eligibility for drought assistance.

Although agricultural losses typically dominate drought impacts, congressional interest in federal drought assistance is not limited to the USDA. For example, the 2012 drought raised interest in whether and to what extent other federal agencies have and are using authorities to assist with managing drought. For example, in addition to operations of federal water resource facilities (discussed below), the U.S. Army Corps of Engineers and the U.S. Bureau of Reclamation also have limited emergency drought authorities and funding; the drought response and recovery authorities for the Corps, Reclamation, and the Farm Service Agency are provided in *Appendix A*. While the Corps and Reclamation authorities have experienced some use, the frequency, impact, and coordination of their use with other federal and state drought efforts have not been monitored and assessed. Most of the Corps drought response and recovery authorities are limited to providing limited and temporary water from its reservoirs for a fee. The exception is its authority to assist with well construction and transport of water to drought-distressed farmers, ranchers, and political subdivisions; the construction costs are a nonfederal responsibility, while the transport costs can be federally funded. Congress has provided Reclamation with a broader set of authorities that allow it to assist with drought response and recovery; these authorities, however, are limited to the 17 western states[71] and Hawaii.

USDA Emergency Assistance: Status and Legislation

Under current U.S. farm policy, financial losses caused by drought and other natural disasters are mitigated primarily by the federal crop insurance program (administered by USDA's Risk Management Agency). From 2000 to 2011, the federal contribution to the crop insurance program averaged about $4.6 billion per year, mostly in premium subsidies. In crop year 2012, when drought adversely affected many growing regions, crop producers were reimbursed for losses totaling $16 billion (as of April 1, 2013). (The total premium subsidy to farmers was nearly $7 billion.)

Since the severe drought of 1988 and until passage of the 2008 farm bill (P.L. 110-246), Congress regularly made supplemental financial assistance available to farmers and ranchers (in addition to crop insurance), primarily in the form of crop disaster payments and emergency livestock assistance. Crop disaster payments, paid to any producer who experienced a major crop loss caused by a natural disaster, totaled $22.34 billion from FY1989 to FY2009.

More recently, under the 2008 farm bill (P.L. 110-246), Congress authorized a $3.8 billion trust fund to supplement crop insurance and cover the cost of making agricultural disaster assistance for losses from droughts and other causes available on an ongoing basis over four years (FY2008-FY2011). Among the programs operating from FY2008 through September 30, 2011, the Livestock Forage Disaster Program (LFP) assisted ranchers who graze livestock on drought-affected pastureland or grazing land. As of March 5, 2013, payments under LFP totaled more than $565 million for losses caused by weather events on or before September 30, 2011.

The expired status of these disaster payment programs received significant congressional attention in 2012 as drought conditions spread across the country. The 112th Congress considered but did not pass omnibus farm legislation, including extension of certain agricultural disaster programs that expired in September 2011. The Senate passed its version of the omnibus 2012 farm bill (S. 3240, the Agriculture Reform, Food, and Jobs Act of 2012) in June 2012. The Senate bill would have retroactively extended the livestock disaster and tree assistance programs, thereby potentially covering losses associated with the 2012 drought. In the House, on July 11, 2012, the House Agriculture Committee passed its farm bill (H.R. 6083, the Federal Agriculture Reform and Risk Management Act of 2012), which included the same combination of disaster programs as in the Senate bill. The bill, however, did not reach the House floor. Other attempts in the 112th Congress to reauthorize and fund the disaster programs were not successful. At the end of the 112th Congress, on January 2, 2013, the five-year 2008 farm bill was extended one year as part of the American Taxpayer Relief Act of 2012 (ATRA; P.L. 112-240). Under ATRA, Congress provided authority to appropriate funds (but no actual funding) for the three livestock programs and the tree assistance program. Neither discretionary funding authority nor resources were provided for the crop disaster program (the Supplemental Revenue Assistance Payments Program or SURE). With farm bill programs expiring in 2013, the 113th Congress may consider the reauthorization of an omnibus farm bill.

In response to the 2012 drought, USDA took a variety of administrative actions. For example, it reduced the interest rate for emergency loans from 3.75% to 2.25%. Also, USDA authorized emergency haying and grazing use of Conservation Reserve Program (CRP) acres for 2012 due to drought conditions.

USDA announced a smaller reduction (10% instead of 25%) on rental payments made to producers on CRP lands used for emergency haying and grazing in 2012. The agency also announced a purchase of up to $170 million in livestock and fish products to help ease the financial effects of drought on meat and catfish producers.

Other federal agencies are also reviewing drought response actions and committing resources or temporarily changing policies that may ease drought burdens. For example, the President in early August 2012 convened a White House Rural Council to review and assess federal agency activities and capabilities. Shortly following the gathering, the Administration announced new measures to address drought impact, as well as a listing of ongoing federal agency efforts to address the 2012 drought.[72] Among the new efforts announced were waivers for federal trucking regulations and additional emergency funding for crop and livestock producers.[73]

Federal Facilities and Drought

Operations of federal water resource facilities, particularly reservoirs behind dams, can both assist in meeting water supply needs during droughts and be vulnerable to droughts. Federal dams, particularly in the West, were constructed in part to provide multi-year storage to help with variations in seasonal and annual precipitation.[74] Sustained hydrological drought nonetheless affects operations of federally managed reservoirs, dams, locks, hydroelectric facilities, and other components of the nation's water infrastructure. For example, numerous Corps reservoirs have drought management plans that result in the curtailing of some benefits (e.g., navigation, hydropower) in order to maintain other benefits (e.g., in-stream flows to support water quality, aquatic species, and river withdrawals for electric power cooling and municipal and industrial water supplies). The Corps' operations of its facilities in the Apalachicola-Chattahoochee-Flint River system during 2007-2008 illustrate management tradeoffs during drought (see box below).

Federal Reservoir Operations During Southeast Droughts

An example of hydrological drought was the 2007-2008 drought in the southeastern United States. A severe drought in the region, beginning with below-average rainfall in spring 2006, exacerbated an ongoing interstate dispute involving Alabama, Florida, and Georgia over water sharing in the Apalachicola-Chattahoochee-Flint (ACF) river system. During the drought, Atlanta's municipal and industrial water users in the upper basin were concerned over the potential loss of their principal water supply, Lake Lanier, a surface water reservoir behind a U.S. Army Corps of Engineers operated dam. Their concern resulted from the decision by the Corps to draw down Lake Lanier in the fall of 2007. The Corps drew down the reservoir to maintain minimum flows in the lower basin Apalachicola River to support species protection, energy production (e.g., power plant cooling), and lower basin municipal withdrawals.

During and after the 2007-2008 drought, additional actions were taken and considered at the state level to manage water demand in droughts. The extent to which similar activities may be implemented or necessary again revives policy questions of what actions should be taken by whom and when in a shared basin to adapt to dry conditions.

Source: NOAA, National Weather Service, Southeast River Forecast Center, When Did the Drought Begin, a Focus on the North Georgia and Atlanta Areas, November 16, 2007.

Similarly, drought conditions in California from 2007 to 2009, coupled with declining fish species, resulted in operational changes to Reclamation facilities, including significantly reduced water deliveries to Central Valley Project contractors, as well as to California's State Water Project (SWP) contractors. Reclamation, whose facilities currently serve over 31 million people in the West and deliver a total of nearly 30 million acre-feet of water[75] annually, faces operational challenges because of conflicts among its water users during drought in states it serves.

For example, severe drought conditions in 2001 in the Klamath River basin, on the Oregon-California border, exacerbated competition for scarce water resources among farmers, Indian tribes, commercial and sport fishermen, other recreationists, federal wildlife refuge managers, environmental groups, and state, local, and tribal governments. Reclamation's decision in April 2001 to withhold water from farmers for in-stream flows for three fish species listed as endangered or threatened under the Endangered Species Act sparked congressional debate that continues today. The Klamath basin again experienced drought conditions in 2010 and again in 2012. Project water flows to Klamath refuges were halted from December 2011 through March 2012. Dry conditions contributed to a cholera outbreak among migrating birds during this time, resulting in the death of thousands of birds that visit the refuges. Early spring precipitation improved hydrological conditions such that Reclamation projected full irrigation deliveries for 2012.[76] However, low lake levels and inflows by April 1, 2013, have resulted in Reclamation postponing spring water deliveries by one to two weeks.[77] Final water deliveries are also affected by flows necessary to protect federally listed Coho salmon and two sucker species. A new biological opinion on project operations through February 2014 is expected in April 2013, which may again affect water deliveries.

Droughts and Navigation:
The 2012-2013 Experience

The dry conditions of 2012 contributed to low flow conditions in portions of the Mississippi River into early 2013. While a full shutdown of barge navigation on the river did not materialize, the risk to navigation and the industries that rely on that navigation (e.g., agriculture, and energy) resulted in intense political interest and the Army Corps of Engineers taking emergency actions (e.g., blasting of rocks in areas with shallow depths) to avoid disruptions.

The situation resulted in increased interest among Mississippi River stakeholders in having federal infrastructure on the Missouri River and its tributaries operated to support Mississippi River navigation; this has prompted introduction of legislative proposals (e.g., S. 565, Mississippi River Navigation Sustainment Act).

Water management issues in the Missouri River basin have also been debated among upper basin and lower basin interests for decades; the intensified interest of Mississippi River stakeholders further complicates the Corps efforts to balance the needs of a diverse mix of basin stakeholders (e.g., agricultural and municipal and industrial water supply users, hydroelectric power users, navigation interests, and species that rely on certain water flows and timing of water releases).

The droughts in California, the Southeast, and the Klamath River basin underscore an underlying difficulty of managing federal reservoirs to meet multipurpose water needs. In the future, the United States might face severe and sustained periods of drought not experienced in the 20th century. If so, disputes over federal infrastructure management like those in California, the ACF basin, and Klamath River basin may increasingly determine short-term actions by Reclamation and the Corps, and result in long-term consequences for congressional oversight and funding.

Drought Forecasts for the United States

Predicting the severity and duration of severe drought over a specific region of the country is not yet possible more than a few months in advance because of the many factors that influence drought. Nevertheless, some modeling studies suggest that a transition to a more arid average climate in the American West, perhaps similar to conditions in precolonial North America, may be underway.[78] Some studies have suggested that human influences on climate, caused by emissions of greenhouse gases, may be responsible for a drying trend; [79] however, other studies appear to indicate an opposite trend or possibility (see above section on "Drought and Climate Change").[80] Whether future greenhouse gas-driven warming can be linked to La Niña-like conditions, or other phenomena related to the El Niño-Southern Oscillation, is unclear.

A likely consequence of higher temperatures in the West would be higher evapotranspiration, reduced precipitation, and decreased spring runoff.[81] These impacts would result from an "acceleration" of the hydrologic cycle, due to increased warming of the atmosphere, which in turn increases the amount of water held in the atmosphere.[82] A possible consequence is more frequent, and perhaps more severe, droughts and floods. However, these changes are unlikely to occur evenly across the United States. Observations of water-related changes over the last century suggest that runoff and streamflow in the Colorado and Columbia River basins has been decreasing, along with the amount of ice in mountain glaciers in the West, and the amount of annual precipitation in the Southwest.[83] Yet the understanding of hydrologic extremes, such as drought, is confounded by other effects such as land cover changes, the operation of dams, irrigation works, extraction of groundwater, and other engineered changes. Forecasting drought conditions at the regional scale, for example for river basins or smaller, is difficult because current climate models are less robust and have higher uncertainty at smaller scales.[84] (For example, see box below on the Colorado River's Lake Mead.)

Even though forecasting drought at the regional scale is difficult, understanding potential changes in long-term trends is important for water managers at all levels—federal, state, local, and tribal. Water project operations and state water allocations are typically based on past long-term hydrological trends; significant deviations from such trends may result in difficult challenges for water managers and water users alike.[85] An example of such a dilemma can be observed in the Colorado River basin.

Colorado River's Lake Mead

A 2008 study asserted that water storage in Lake Mead has a 50% probability by 2021 to "run dry" and a 10% chance by 2014 to drop below levels needed to provide hydroelectric power under current climate conditions and without changes to water allocation in the basin. This study raised awareness of the vulnerability of western water systems but drew criticism that global climate models are insufficient to forecast climate change effects at the regional scale. Some western water officials were especially critical of the report's assertions. One explained that Reclamation and other agencies had recently developed new criteria for the allocation of Colorado River water in times of shortages (shortage criteria), including drought, and commented that the likelihood that Lake Mead would run dry was "absurd." The study was based on predictions of future warming in the West without increased precipitation.

In a 2009 follow-up study, the same authors acknowledged that the ability of the Colorado River system to mitigate drought could be managed if the users found a way to reduce average deliveries, thereby maintaining water levels in Lake Mead and Lake Powell at consistently higher elevations. Maintaining higher water levels would increase the capacity of the Colorado River system to buffer itself against low precipitation years. Even so, the authors noted, global climate models are in broad agreement that the southwestern United States is likely to become warmer and more arid, especially in the Colorado River drainage basin. In addition, paleoclimate studies suggest that the 20th century was the wettest or second-wettest century for at least 500 years and possibly over the past 1,200 years. Notwithstanding potential climate change, the paleoclimate data suggest that average future precipitation in the Colorado River basin is unlikely to match what hydrologists believe were relatively wet 20th-century levels, and upon which water allocation decisions were made.

In December 2012, the Bureau of Reclamation and basin state agencies released the Colorado River Basin Water Supply and Demand Study, which was undertaken to address uncertainties over Colorado River water supplies. The study assessed future water supply and demand imbalances under several different supply and demand scenarios in order to provide background information and data for decision makers. Although the study also solicited water supply and demand management solutions, it did not provide recommendations for future management. Key findings echoed concerns noted in earlier studies. Notably, the study confirmed that the Colorado River is over-allocated given existing and projected water supply levels and that "the long-term projected imbalance in future supply and demand [based on median projections of each] is about 3.2 million acre-feet by 2060." The authors note that this figure does not take into account reservoir storage and management and that factors contributing to the supply and demand projections are highly variable. This disparity makes planning even more difficult for basin water managers.

Management of reservoir operations and future agreements among basin states will continue to play a large role in the system's viability. Although the system was able to meet all downstream water requests in recent years despite experiencing the worst drought in a century, there is no guarantee that a longer-term drought or successive low water years will not become the new norm. That said, the current system has enormous capacity to recover if there are above normal water years given its 60 million acre-feet storage capacity—a resilience that was demonstrated over the recent 11-year drought.

Sources: Tim P. Barnett and David W. Pierce, "When Will Lake Mead Go Dry?" Water Resources Research, vol. 44 (March 29, 2008), p. W03201, DOI:10.1029/2007WR006704; Felicity Barringer, "Lake Mead could be within a few years of going dry, study finds," New York Times (February 13, 2008); Jenny Dennis, "Stunned Scientists: 'When Will Lake Mead Go Dry?'" Rim Country Gazette (February 28, 2008), quoting Larry Dozier, Central Arizona Project deputy general manager; Timothy P. Barnett and David W. Pierce, "Sustainable Water Deliveries from the Colorado River in a Changing Climate," Proceedings of the National Academy of Sciences, vol. 106, no. 18 (May 6, 2009); U.S. Dept. of the Interior, Bureau of Reclamation, Colorado River Basin Water Supply and Demand Study (December 2012).

Conditions in the Colorado River basin over the last decade, including recent low reservoir levels in Lake Mead and Lake Powell, and low flows in the Upper Basin, raise the issue of what is the baseline for average hydrologic cycles now and in the future. The allocation of Colorado River water supplies was agreed upon by lower and upper basin states in the early part of the 20th century based on hydrologic data from what scientists now know was a relatively wet period in the history of the Colorado River basin.[86] If long-term reduced runoff predictions for the basin are borne out (see box above on Colorado River's Lake Mead and earlier section on "Conditions in the Colorado River Basin"), then water allocation policies for regions like the Colorado River basin may again need to be revisited.[87] In the meantime, Colorado River basin states have negotiated "shortage criteria" and "interim guidelines" for managing Colorado River water supplies during times of shortages.[88] Additionally, Reclamation in December 2012 released a new Colorado River basin supply and demand study, which examined several different future water supply and demand scenarios.[89]

POLICY CHALLENGES

Severe drought can cause significant economic harm, affect nearly all areas of the country, and exacerbate water competition. Nonetheless, several key factors make comprehensive drought policy at the national level a challenge, including:

- the gradual, or "creeping," nature of drought;
- split federal and nonfederal drought response and management responsibilities;
- a patchwork of federal programs and oversight with little coordination; and
- differences in regional conditions and drought risk in terms of the drought hazard, vulnerability, and potential consequences.

Drought conditions often develop slowly, are not easily identified initially, and are challenging to forecast beyond 30 to 90 days. Consequently, drought declarations are made well after onset— typically once impacts are felt. This situation makes it difficult to mitigate or prevent drought impacts. Further, even though drought generally is continuously occurring somewhere in the United States, the unpredictability of its location, duration, and severity complicates preparation for implementation of responses.

When severe meteorological drought affects a region, the supply of available water often shrinks before use is reduced. Adjusting down the use of water as drought persists and supplies shrink can be difficult. Actually, droughts can increase demand on water supplies (e.g., lower soil moisture results in increased demand for irrigation and landscape watering). The lack of flexibility of existing water access and use arrangements sometimes limits the scope and speed of some drought responses. Federal, state, and local authorities make water resource decisions within the context of multiple and often conflicting laws and objectives, competing legal decisions, and entrenched institutional mechanisms, including century-old water rights and long-standing contractual obligations (i.e., long-term water delivery and power contracts). Typically, how access to and competition for water is managed (e.g., permitting of water withdrawals) and how reductions in water supply are managed (e.g., shared reductions under a riparian system of water rights versus reductions based on the priority in time of a water right) is determined by state law and at times through interstate compacts. Additionally, state and local laws can determine how easily water can be transferred among users. These access, reduction, and transfer arrangements can significantly affect the behavior, incentives, and opportunities available to water users during droughts. Fundamental changes to the access, reduction, and transfer arrangements are largely outside of the realm of federal action, and are largely determined by each state; however, such activities might be directly or indirectly influenced by changes in federal policies and activities.

A mismatch between supply and demand during droughts underscores the responsibility of stakeholders to anticipate the influence of drought and plan and act accordingly. The federal government has several drought monitoring and response programs. While drought planning and mitigation responsibilities lie largely at the state and local level, the federal government also provides some drought planning assistance. Much of this assistance historically has been concentrated on the West. Additionally, the federal government often provides emergency funding for drought relief that is primarily aimed at easing the economic impacts. The National Drought Commission and others have noted, however, that federal relief programs and emergency funding provide little incentive for state and local planning and efforts to reduce social and economic vulnerability to drought. A policy issue particularly relevant to state and local decision makers is the role and types of demand management tools to employ during a drought (e.g., lawn watering restrictions, incentives to curtail irrigation during droughts, scarcity pricing). How a state distributes and administers its waters among competing uses can affect what drought response tools are available to it and to water users.

A further challenge is lack of a cohesive national drought policy at the federal level, and lack of a lead agency coordinating federal programs. Rather, several federal programs have been developed over the years, often in response to specific droughts. Additionally, occasional widespread economic effects have prompted creation of several federal relief programs. These programs are overseen by different congressional committees. Whether this fragmentation results in duplication, waste, and gaps, or whether it reflects the complexity of preparing and responding to drought and the different responses needed by a wide range of stakeholders (e.g., irrigated agriculture, dry land farming, municipal water utilities) is part of the debate about how to proceed with cost-effective management of the nation's drought risk and who bears the consequences of drought. (See box below for an example of how water access and transfer arrangements played a significant role in shaping Australia's drought resilience and adaptation.)

The 2012 drought, like other recent severe weather events, has contributed to the ongoing public and policy discussion about the influence of human actions on climate and the future of U.S. actions on climate change mitigation.

Australia's Drought Experience:Water Markets as Drought Management

Australia experienced a historic drought from 1997 to 2009, known as the Millennium Drought. The drought tested a preexisting multi-pronged national water reform initiative; one aspect of the reform was the development of water markets. To develop water markets, the initiative had promoted reform of state law to clarify the property right associated with a water right and facilitated the means to buy and sell perpetual water rights and short-term allocations in basins that were fully allocated.

Water trade increased significantly during the later years of the drought as allocations fell and markets matured. Allocations in some sub-basins during the worst of the drought reached as low as 20% of a full allocation. While gross domestic production dropped by $2 billion-$3 billion in Australia's most significant agricultural basin during each of the worst drought years, the ability to trade water is estimated to have reduced losses by roughly $1 billion during each drought year.

The market's ability to move scarce water to uses with higher economic value is credited with assisting Australia's rural economy to ride out the drought as well as it did by getting more value per unit of water used.

For example, some dairy farmers sold their water rights and purchased fodder, rather than growing it themselves.

Agricultural businesses increasingly used buying and selling in the water market as a coping mechanism as the drought persisted. With water availability high in many basins since 2009, market water prices have fallen, and rice and cotton production, which had declined during the drought, have picked back up.

Water markets were not established in Australia without controversy and criticism. While not solely responsible, water rights trading contributed to trends producing significant economic adjustment, particularly in rural agricultural communities. Nonetheless, contemplating the consequences for Australia, especially its agricultural communities and businesses, of such a severe drought under a less flexible water rights regime has increased internal support for the use and further improvement of water markets.

The broader water reform initiative produced some disappointments, as well as successes. The broader reform is criticized for falling short of achieving ecologically sustainable levels of surface water withdrawals. Consequently, the recent discourse about the next steps in Australian water policy has focused on how to establish sustainable levels of withdrawals that can maintain ecosystems and support regional economies and how to cost-effectively secure the water for the environment.

Australia's government uses the water markets to transition water out of existing uses for use in meeting environmental flow goals; to date, the Australian government's purchase of water rights for the environment using the market has been less expensive than obtaining the water through infrastructure efficiency improvements.

Source: National Water Commission (Australia), The Impacts of Water Trading in the Southern Murray-Darling Basin Between 2006-07 and 2010-11, April 2012.

LEGISLATIVE ACTION

Congress has long recognized the lack of coordinated drought planning and mitigation activities among federal agencies and the predominance of a crisis management approach to dealing with drought. Over the last 15 years, legislative action has focused on the question of whether there is a need for a national drought policy. For example, in 1998, Congress passed the National Drought Policy Act (P.L. 105-199), which created the National Drought Policy Commission. In 2000, the commission submitted to Congress a comprehensive report that included policy recommendations. Congress has considered recommendations from the commission's 2000 report; to date, it has enacted one part of the recommendations (the National Integrated Drought Information System, discussed below). Congress also considered, but did not enact, legislation creating a National Drought Council during deliberations on the 2008 farm bill. Recent congressional deliberations focused on the 2012 reauthorization of the farm bill, which was given a one-year extension (through 2013) in January 2013, although other legislation is also pending.[90]

The commission findings, the proposed council, and the 2008 and 2013 farm bills are discussed below.

The National Drought Policy Act of 1998

In passing the National Drought Policy Act of 1998, Congress found that "at the Federal level, even though historically there have been frequent, significant droughts of national consequences, drought is addressed mainly through special legislation and ad hoc action rather than through a systematic and permanent process as occurs with other natural disasters."[91] Further, Congress found an increasing need at the federal level to emphasize preparedness, mitigation, and risk management. Those findings are consistent with a recognition of the inevitability, albeit unpredictability, of severe drought occurring. The act created the National Drought Policy Commission, and required it to conduct a study and report to Congress on

- what is needed to respond to drought emergencies;
- what federal laws and programs address drought;
- what are the pertinent state, tribal, and local laws; and
- how various needs, laws, and programs can be better integrated while recognizing the primacy of states to control water through state law.

In May 2000, the commission submitted its report.[92] The commission concluded that the United States needed to embrace a national drought policy with preparedness at its core. It recommended that Congress enact a National Drought Preparedness Act, which would establish a federal-nonfederal partnership through a National Drought Council. The council would function to further national drought policy goals. The commission's report provided 29 specific recommendations to achieve the goals of national drought policy. Most of the specific recommendations were targeted at the President and federal agencies, coupled with calls for Congress to fund drought-related activities in support of the recommendations. (*Appendix B* of this report lists the five goals and illustrative recommendations from the commission's report.) As background for its recommendations, the commission noted the patchwork nature of drought programs, and that despite a major federal role in responding to drought, no single federal agency leads or coordinates drought programs—instead, the federal role is more of "crisis management."[93]

National Drought Preparedness Legislation and the 2008 Farm Bill

National Drought Preparedness Act bills were introduced in 2002 (107th Congress), 2003 (108th Congress), and 2005 (109th Congress), but were not enacted. Similar stand-alone legislation was introduced in the 110th Congress; however, the House-passed version of H.R. 2419, the Farm, Nutrition, and Bioenergy Act of 2008 (also known as the 2008 farm bill), contained a section creating a National Drought Council. This section of the 2008 farm bill would have charged the council with creating a national drought policy action plan, which would have incorporated many of the components recommended in the commission's report; however, it was not included in the conference agreement. Although the Senate version of H.R. 2419 did not contain a similar section, the Senate bill authorized permanent disaster payments in hopes of precluding the need for ad hoc disaster payments. The conference agreement on the 2008 farm bill (P.L. 110-246, enacted June 18, 2008) included a new $3.8 billion trust fund to cover the cost of making agricultural disaster assistance available on an ongoing basis over the following four years. The assistance was available for disasters occurring on or before September 30, 2011. In January 2013, as part of a one-year extension of the 2008 farm bill, four of the five disaster programs were reauthorized through FY2013, but Congress provided no funding. With farm bill programs expiring in 2013, the 113th Congress is expected to consider the reauthorization of an omnibus farm bill, including agricultural disaster programs (see box on page 18).

National Integrated Drought Information System

Although Congress has not enacted comprehensive national drought preparedness legislation, it acted on the second of five commission goals by passing the National Integrated Drought Information System (NIDIS) Act of 2006 (P.L. 109-430). That goal called for enhanced observation networks, monitoring, prediction, and information delivery of drought information. P.L. 109-430 established NIDIS within the National Oceanic and Atmospheric Administration (NOAA) to improve drought monitoring and forecasting abilities.[94] The

NIDIS authorization expired in 2012. A NIDIS reauthorization bill, S. 376, has been introduced in the 113th Congress.

CONCLUSION

Drought is a natural hazard with potentially significant economic, social, and ecological consequences. History suggests that severe and extended droughts are inevitable and part of natural climate cycles. Drought has for centuries shaped the societies of North America and will continue to do so into the future. Current understanding is that the physical conditions causing drought in the United States are linked to sea surface temperatures in the tropical Pacific Ocean. For example, the 2011 severe drought in Texas is thought to be linked to La Niña conditions in the Pacific Ocean. Increasingly, studies are projecting the long-term role that droughts may play in regional climate patterns. Nonetheless, available technology and science remains limited to forecasting specific drought a few months in advance for a region. The prospect of extended droughts and more arid baseline conditions in parts of the United States represents a challenge to existing public policy responses for preparing and responding to drought, and to federal water resource projects in particular, because their construction was based largely on 19th- and 20th-century hydrologic conditions.

Over time, Congress has created various drought programs, often in response to specific droughts and authored by different committees. Crafting a broad drought policy that might encompass the jurisdiction of many different congressional committees is often difficult. Additionally, although many water allocation and other water management responsibilities largely lie at the state or local level, localities and individuals often look to the federal government for relief when disasters occur. This is similar to the situation for flood policy, and water policy in general, at the national level. The National Drought Policy Commission recognized these patterns, and they underlie many of its recommendations to Congress.[95] The currently fragmented approach can be costly to national taxpayers; however, it is not certain that increased federal investment (especially vis-à-vis the potential for tailored local and state investment) in drought preparation, mitigation, and improved coordination would produce more economically efficient outcomes.

The overall costs to the federal government and the nation as a result of extreme drought, apart from relief to the agricultural sector, are difficult to assess in part because of the broad nature of drought's impacts. Drought can result in water restrictions affecting municipal and industrial users, decreased hydropower generation and power plant cooling efficiency, navigation limitations and disruptions, harm to drought-sensitive species (benefits to other species), and increased fire risk, among other effects.

Congress may opt to revisit the commission's recommendations and reevaluate whether current federal practices could be supplemented with actions to coordinate, prepare for, and respond to the unpredictable but inevitable occurrence of drought. Given the daunting task of managing drought, Congress also may consider proposals to manage drought impacts, such as assisting localities, industries, and agriculture with developing or augmenting water supplies. This could take multiple forms: construction or permitting of reservoirs, the reallocation of water supplies at existing facilities, promotion of alternative water sources (e.g., reuse, desalination), or water conservation and efficiency. Congress also may move to examine how the two major federal water management agencies, the Corps and Reclamation, plan for and respond to severe drought and account for its impacts.

APPENDIX A. DROUGHT-RELATED FEDERAL RESPONSE AND RECOVERY AUTHORITIES

Table A-1. Drought-Related Federal Response and Recovery Authorities Army Corps of Engineers, Bureau of Reclamation, and Farm Service Agency (not including USDA crop insurance programs)

Agency	Program Name and Brief Description	Primary Authorities
Army Corps of Engineers	Temporary water withdrawal, for a fee, from Corps facilities during drought for municipal and industrial purposes. Administration policy limits application to governor-declared drought emergencies and to 99 acre-feet.	Flood Control Act of 1944, as amended (33 U.S.C. 708)
Army Corps of Engineers	Construction of wells for and transport of water to drought-distressed farmers, ranchers, and political subdivisions. Non-transport costs are nonfederal.	Disaster Relief Act of 1974 (33 U.S.C. 701n)
Army Corps of Engineers	Contracts for temporary sale of surplus water for municipal and industrial use from Corps reservoirs at a fee determined by the Secretary of the Army.	Flood Control Act of 1944 (33 U.S.C. 708)
Army Corps of Engineers	Interim use for irrigation of un-contracted water designated for municipal and industrial use at Corps reservoirs for a fee.	Water Resources Development Act of 1986 (43 U.S.C. 390)
Bureau of Reclamation (17 Western States and Hawaii)	Temporary construction, management, and conservation activities (well construction can be permanent but must be designed to address drought impacts). Costs incurred by Reclamation performing or contracting for the work are federal (nonreimbursable); other costs are to be repaid to the federal government (reimbursable).	Reclamation States Emergency Drought Relief Act of 1991, as amended (43 U.S.C. 2201, et seq.)a
Bureau of Reclamation (17 Western States and Hawaii)	Loans with interest to water users (that demonstrate an ability to pay within the terms of the loan) for construction of temporary facilities and permanent wells, conservation activities, and acquisition of water to mitigate droughts.	Reclamation States Emergency Drought Relief Act of 1991, as amended (43 U.S.C. 2201, et seq.)[a]
Bureau of Reclamation(17 Western States and Hawaii)	Authorizes Reclamation to facilitate water purchases and transfers through nonfinancial assistance, and to provide financial assistance for contingency plans. Cost for purchasing, storing, and conveying water are generally nonfederal (reimbursable).	Reclamation States Emergency Drought Relief Act of 1991, as amended (43 U.S.C. 2201, et seq.)a
Bureau of Reclamation (17 Western States and Hawaii)	Authorizes Reclamation purchases of water from its project contractors and sold under temporary contracts at full cost recovery. Authorizes Reclamation's participation in state water banks.	Reclamation States Emergency Drought Relief Act of 1991, as amended (43 U.S.C. 2201, et seq.)a
Bureau of Reclamation (17 Western States and Hawaii)	Temporary procurement (not to exceed two years) of Reclamation project and non-project water for managing fish and wildlife impacts due to drought or the operation of a Reclamation project during drought	Reclamation States Emergency Drought Relief Act of 1991, as amended (43 U.S.C. 2212)

Table A-1. (Continued)

Agency	Program Name and Brief Description	Primary Authorities
	conditions. Price charged for water is to recover the federal costs of the Reclamation facilities.	
Bureau of Reclamation	Emergency fund established to assure continuous operation of Reclamation facilities. Also, other site or drought specific authority.	Act of June 26, 1948, as amended (43 U.S.C. 502) P.L. 95-18, as amended (43 U.S.C. 502 note)
Farm Service Agency	Noninsured Assistance Payments (NAP)- provides direct payments to crop producers who experience a significant crop loss, but are not eligible for federal crop insurance.	7 U,S,C, 1501 et seq. and U.S.C. 1509
Farm Service Agency	Emergency Loans- low-interest government loans to producers suffering from production and physical losses located in or adjacent to a county that has been declared a disaster by the President or USDA Secretary.	7 U.S.C. 1961 et seq.
Farm Service Agency	Livestock Forage Program (LFP) provides financial assistance to producers who suffered grazing losses due to drought or fire between January 1, 2008, and September 30, 2011. Authorization is extended (but not funded) through September 30, 2013.	7 U.S.C. 1531
Farm Service Agency	Livestock Indemnity Program (LIP) compensates ranchers at a rate of 75% of market value for livestock mortality caused by a disaster between January 1, 2008, and September 30, 2011. Authorization is extended (but not funded) through September 30, 2013.	7 U.S.C. 1531
Farm Service Agency	Tree Assistance Program (TAP) provides financial assistance to qualifying nursery tree growers and orchardists to replant or rehabilitate eligible trees, bushes and vines damaged by natural disasters, including floods, occurring on or after January 1, 2008, and before October 1, 2011. Authorization is extended (but not funded) through September 30, 2013.	7 U.S.C. 1531
Farm Service Agency	Emergency Assistance for Livestock, Honey Bees, and Farm-Raised Fish Program (ELAP) compensates producers for disaster losses not covered under other disaster programs. Authorization is extended (but not funded) through September 30, 2013.	7 U.S.C. 1531
Farm Service Agency	Emergency Conservation Program (ECP)- Provides emergency funding and technical assistance to producers to rehabilitate farmland damaged by natural disasters, including implementing emergency water conservation measures in response to severe droughts.	16 U.S.C. 2201-2205
Farm Service Agency	Emergency Forest Restoration Program (EFRP) provides payments to eligible owners of nonindustrial private forest (NIPF) land in order to carry out emergency measures to restore land damaged by a natural disaster, including drought.	16 U.S.C. 2206

Source: CRS.

Notes: For a discussion of USDA crop insurance programs see CRS Report R40532, Federal Crop Insurance: Background , by Dennis A. Shields.

[a] For the draft agency policy on implementation, see Bureau of Reclamation, Draft Reclamation Manual: Directives and Standards, Title I, Reclamation States Emergency Drought Relief Act of 1991, as amended, 2002, http://www.usbr.gov/drought

APPENDIX B. EXCERPTS FROM THE 2000 NATIONAL DROUGHT POLICY COMMISSION REPORT TO CONGRESS

In 2000, the National Drought Policy Commission published its report to Congress, *Preparing for Drought in the 21st Century—A Report of the National Drought Policy Commission.*[96] The commission concluded that the United States needed to embrace a national drought policy with preparedness at its core. It recommended that Congress enact a National Drought Preparedness Act, which would establish a federal-nonfederal partnership through a National Drought Council. The council would function to further national drought policy goals. The council's recommended duties included promoting cooperation and effective delivery of drought programs, evaluation of regional needs and opportunities, and assessments of drought-related assistance and initiatives, and post-drought impact assessments.

The commission also recommended five broad national policy goals and a number of recommendations for furthering each of the goals. Below are the five goals and below those are some examples of the types of recommendations accompanying each of the goals; many more recommendations are proposed in the commission's report. To date, Congress has enacted legislation furthering Goal 2 on improved observation. Congress also has taken action on insurance and financial strategies; their effect on improving drought preparedness (i.e., reducing drought vulnerability) versus increasing the relief for the impacts of drought has not been evaluated. This type of assessment potentially would fall under the duties of a National Drought Council as recommended by the commission.

Policy Statement[97]

- Favor preparedness over insurance, insurance over relief, and incentives over regulation.
- Set research priorities based on the potential of the research results to reduce drought impacts.
- Coordinate the delivery of federal services through cooperation and collaboration with nonfederal entities.

Goals[98]

Goal 1. Incorporate planning, implementation of plans and proactive mitigation measures, risk management, resource stewardship, environmental considerations, and public education as the key elements of effective national drought policy.

Goal 2. Improve collaboration among scientists and managers to enhance the effectiveness of observation networks, monitoring, prediction, information delivery, and applied research and to foster public understanding of and preparedness for drought.

Goal 3. Develop and incorporate comprehensive insurance and financial strategies into drought preparedness plans.

Goal 4. Maintain a safety net of emergency relief that emphasizes sound stewardship of natural resources and self-help.

Goal 5. Coordinate drought programs and response effectively, efficiently, and in a customer-oriented manner.

Illustrative Recommendations for Goal 1: Planning and Mitigation[99]

- Congress should adequately fund existing drought preparedness programs.
- President should direct appropriate federal agencies to effectively meet the drought planning needs of those areas not traditionally served (e.g., eastern United States).
- The President should direct all appropriate federal agencies to study their programs for potential impacts on drought.
- The president should direct all appropriate federal agencies to develop and implement drought management plans for federal facilities (e.g., military bases, federal office complexes, federal prisons).

End Notes

[1] Edward R. Cook, et al., "North American drought: reconstructions, causes, and consequences," Earth-Science Reviews, vol. 81 (2007): pp. 93-134. Hereafter referred to as Cook et al., 2007.

[2] Donald A. Wilhite, et al., Managing Drought: A Roadmap for Change in the United States (Boulder, CO: The Geological Society of America, 2007), p. 12; at http://www.geosociety.org/meetings/06drought/roadmap.pdf.

[3] These are the categories used by the National Drought Mitigation Center (NDMC). The NDMC helps prepare the U.S. Drought Monitor and maintains its website.

[4] NDMC data collected since 2000. U.S. Drought Monitor at the NDMC in Lincoln, NE. See http://drought monitor.unl.edu/DM_tables.htm?archive.

[5] Extreme drought is the fourth of five categories indicating drought conditions, ranging from abnormally dry to exceptional drought, according to the National Drought Mitigation Center.

[6] In some years or months, however, no part of the country was under extreme or exceptional drought. For example, from January 2000 through early April 2000, extreme or exceptional drought did not affect any portion of the country.

[7] For the contiguous United States, 80% of the land area was affected by abnormally dry or drought conditions. U.S. Drought Monitor, http://droughtmonitor.unl.edu/DM_tables.htm?conus.

[8] U.S. Drought Monitor, Drought Monitor Archive Tables, http://droughtmonitor.unl.edu/dmtabs_archive.htm.

[9] Martin Hoerling et al., An Interpretation of the Origins of the 2012 Central Great Plains Drought, National Oceanic and Atmospheric Administration, Assessment Report: NOAA Drought Task Force Narrative Team, March 20, 2013, ftp://ftp.oar.noaa.gov/CPO/pdf/mapp/reports/2012-Drought-Interpretation-final.web-041113.pdf.

[10] Ibid., p.1.

[11] Ibid., p. 4.

[12] Hoerling et al., 2013, p. 10.

[13] Ibid., p.22.

[14] NDMC, http://www.drought

[15] Evapotranspiration may be defined as the loss of water from a land area through transpiration from plants and evaporation from the soil and surface water bodies such as lakes, ponds, and manmade reservoirs.

[16] Permanently arid conditions reflect the climate of the region, which is the composite of the day-to-day weather over a longer period of time. Climatologists traditionally interpret climate as the 30-year average. See NDMC, http://www.drought.unl.edu/DroughtBasics/WhatisClimatology.aspx.

[17] For more on fire, see CRS Report RL30755, Forest Fire/Wildfire Protection, by Kelsi Bracmort.

[18] NOAA, National Climatic Data Center, http://www.ncdc.noaa.gov/oa/climate

[19] Ibid.

[20] The complete designations are referred to as experimental objective blends of drought indicators, http://www.cpc.ncep.noaa.gov/products/predictions/tools/edb/droughtblends.php.

[21] For example, the Palmer Drought Index has been widely used by the U.S. Department of Agriculture to determine when to grant emergency drought assistance. See NDMC, http://drought.unl.edu/ Planning/Monitoring/ ComparisonofIndicesIntro/PDSI.aspx.

[22] The five key indicators include the Palmer Drought Index, the Climate Prediction Center soil moisture model, U.S. Geological Survey weekly streamflow data, the Standardized Precipitation Index, and short- and long-term drought indicator blends. For a discussion of drought indices, see NDMC, http://droughtmonitor.unl.edu/current.html.

[23] U.S. Drought Monitor, http://www.drought

[24] The "S" and "L" terms in Figure 1 give information on the nature of the drought in the affected region. For more information on the reasoning behind the classification schemes, see http://droughtmonitor.unl.edu/classify.htm.

[25] U.S. Drought Monitor, http://droughtmonitor.unl.edu/DM_tables.htm?TX.

[26] See the U.S. Drought Monitor, Texas, on October 4, 2011, http://droughtmonitor.unl.edu/archive.html.

[27] "Climate Abyss: Weather and Climate Issues with John Nielsen-Gammon," Texas Drought Update, March 23, 2012, http://blog.chron.com/climateabyss/2012/03/texas-drought-update/.

[28] U.S. Drought Monitor, Drought Condition (Percent Area): Texas, http://droughtmonitor.unl.edu/ DM_tables.htm?TX.

[29] Office of the Texas State Climatologist, "Texas Drought Officially the Worst Ever," August 4, 2011, http://tamunews.tamu.edu/2011/08/04/texas-drought-officially-the-worst-ever/.

[30] John W. Nielsen-Gammon, The 2011 Texas Drought: A Briefing Packet for the Texas Legislature, October 31, 2011, p. 29, http://climatexas.tamu.edu/files/2011_drought.pdf.

[31] Office of the Texas State Climatologist, "Texas Drought Officially the Worst Ever," August 4, 2011, http://tamunews.tamu.edu/2011/08/04/texas-drought-officially-the-worst-ever/.

[32] For more information about the hydrology and policy issues involved in the 2007-2009 California drought, see CRS Report R40979, California Drought: Hydrological and Regulatory Water Supply Issues, by Betsy A. Cody, Peter Folger, and Cynthia Brougher.

[33] California Department of Water Resources, California's Drought of 2007-2009—An Overview, September 2010, http://www.water

[34] http://cdec.water

[35] Office of Governor Edmund G. Brown, Jr., "A Proclamation by the Governor of the State of California—Drought," http://gov.ca.gov/news.php?id=16997.

[36] For information on current water supply conditions and historical water allocations to federal water contractors in California, see http://www.usbr.gov/mp/PA/water/.

[37] The Delta smelt is a species of fish listed as threatened under the federal Endangered Species Act and as endangered under the California Endangered Species Act. Natural Resources Defense Council v. Kempthorne, No. 1:05-cv-1207 OWW GSA (E.D. Cal., December 14, 2007).

[38] See http://www.usbr.gov/mp/cvo/vungvari/water

[39] U.S. Department of the Interior, Bureau of Reclamation, Colorado River Basin Water Supply and Demand Study, December, 2012, p. SR-3.

[40] See description at http://www.usbr.gov/uc/water/crsp/cs/gcd.html. (Accessed April 17, 2013.)

[41] For more information on the Law of the River, see http://www.usbr.gov/lc/region/g1000/lawofrvr.html.

[42] See NDMC, at http://drought

[43] Ibid.

[44] Ibid.

[45] Cook et al., 2007.

[46] Hoerling, Martin and Arun Kumar, "The perfect ocean for drought," Science, vol. 299 (January 31, 2003), pp. 691- 694. Hereafter referred to as Hoerling and Kumar, 2003.

[47] Herweiger, Celine, Richard Seager, and Edward Cook, "North American droughts of the mid to late nineteenth century: a history, simulation and implication for Mediaeval drought," The Holocene, vol. 15, no. 2 (January 31, 2006), pp. 159-171. Hereafter referred to as Herweiger et al., 2006.

[48] Cook et al., 2007.

[49] Richard Seager et al., "Model projections of an imminent transition to a more arid climate in southwestern North America," Science, vol. 316 (May 25, 2007): pp. 1181-1184.

[50] Hoerling and Kumar, 2003.

[51] Cook et al., 2007.

[52] Proxies are indirect measurements typically used where direct measurements are unavailable. Tree rings can be used as a proxy for measuring dryness and drought. Similarly, ice cores from glaciers and polar caps can be used as proxies for measuring atmospheric temperatures and carbon dioxide concentrations from thousands of years ago.

[53] Cook et al., 2007.

[54] Herweiger et al., 2006.

[55] Ibid.

[56] The Carey Act, signed into law on August 18, 1894 (Chapter 301, Section 4, 28 Stat. 422), initially made available up to 1 million acres of federal land in each state, provided that the state met several requirements for the eventual development of water resources for reclamation. Some observers have suggested that the failure of the Carey Act to foster irrigation projects in all the land made available, compounded in part by the 1890-1896 drought, led to the Reclamation Act of 1902 and the emergence of the Bureau of Reclamation in the 20th century. (See Marc Reisner, Cadillac Desert (New York, New York, Penguin Books, 1986)).

[57] Fye, F., D.W. Stahle, and E.R. Cook, "Paleoclimate analogs to twentieth century moisture regimes across the United States," Bulletin of the American Meteorological Society, 2003, vol. 84, pp. 901-909.

[58] For example, one report showed that 42% of the area studied in the American West was affected by drought during the years 900 to 1300, versus 30% between 1900 and 2003, a 29% reduction in the average area affected by drought between the two periods. See Cook et al., 2007.

[59] A discussion of ENSO is provided elsewhere in this report.

[60] S. D. Solomon et al., "Contribution of Working Group I to the Fourth Assessment Report of the Intergovernmental Panel on Climate Change," 2007, Cambridge University Press, Cambridge, United Kingdong and New York, NY.

[61] C. B. Field et al., IPCC, "Managing the Risks of Extreme Events and Disasters to Advance Climate Change Adaptation," 2012, Cambridge University Press, Cambridge, United Kingdom and New York, NY, p. 170.

[62] C.B. Field et al., IPCC, 2012, p. 170.

[63] According to the report, confidence in the validity of a finding is based on the type, amount, quality, and consistency of evidence and on the degree of agreement. Confidence is expressed qualitatively: low, medium, high.

[64] C.B . Field et al., IPCC, 2012, p. 172.

[65] Quirin Schiermeier, "Climate Models Fail to 'Predict' US Droughts," Nature, vol. 496, no. 7445 (April 16, 2013), http://www.nature.com/news/climate

[66] For more information, see http://drought.unl.edu/Planning/PlanningInfobyState.aspx. [67]http://www.fsa.usda.gov/FSA/newsReleases?area=newsroom&subject=landing&topic=pfs&newstype=prfactsheet&type=detail&item=pf_20120720_insup_en_ed_desigp.html

[68]http://usda.gov/wps/portal/usda/usdahome?contentid=2012/08/0271.xml&navid=NEWS_RELEASE&navtype=RT& parentnav=LATEST_RELEASES&edeployment_action=retrievecontent

[69] See the USDA FSA crop disaster map (accessed April 17, 2013); http://www.fsa.usda.gov/Internet/FSA_File/disaster_map_cropyr_2013.pdf

[70] For more information about the Stafford Act, see CRS Report RL33053, Federal Stafford Act Disaster Assistance: Presidential Declarations, Eligible Activities, and Funding, by Francis X. McCarthy; and CRS Report R41981, Congressional Primer on Major Disasters and Emergencies, by Francis X. McCarthy and Jared T. Brown.

[71] Per the Reclamation Act of 1902, as amended, the 17 Western states often referred to as "Reclamation States" include the following: Arizona, California, Colorado, Idaho, Kansas, Montana, Nebraska, Nevada, New Mexico, North Dakota, Oklahoma, Oregon, South Dakota, Texas, Utah, Washington, and Wyoming.

[72] See http://www.whitehouse.gov/the-press-office/2012/08/07/fact-sheet-president-

[73] Ibid.

[74] Reclamation is a central player in water resource management in the West. A devastating drought at the end of the 19th century probably was a factor leading to the 1902 Reclamation Act that launched the federal

reclamation effort and Reclamation itself. See Marc Reisner, Cadillac Desert (New York: Penguin, 1986), pp. 108-109. Other research suggests that failures of some late 19th century private irrigation projects, undertaken following passage of the Carey Act (see footnote 56), may have occurred in part due to drought conditions.

[75] One acre-foot is enough water to cover one acre of land one foot deep. An acre-foot is equivalent to 325,851 gallons.

[76] U.S. Bureau of Reclamation, Klamath Project 2012 Operations Plan, April 6, 2012, p. 5, http://www.usbr.gov/mp/ kbao/docs/summer_operations.pdf.

[77] U.S. Bureau of Reclamation, press release, Reclamation Announces Delay in Water Deliveries for Klamath Project, April 2, 2013.

[78] Richard Seager et al., "Model projections of an imminent transition to a more arid climate in southwestern North America," Science, vol. 316 (May 25, 2007), pp. 1181-1184.

[79] Tim P. Barnett, et al., "Human-induced changes in the hydrology of the western United States," Science, vol. 319 (February 22, 2008), pp. 1080-1082.

[80] C.B. Field, et al., IPCC, 2012, p. 170.

[81] Research results are emerging, however, that suggest that local and regional patterns of precipitation may be variable, and parts of a region or a state could receive higher precipitation than the current average, even if the overall trend over the broader area is towards less precipitation. See K. T. Redmond, "Climate Change in the Western United States: Projections and Observations," Eos Trans. AGU, 90(52), Fall Meet. Suppl., Abstract U11D-02, 2009.

[82] National Research Council, Committee on Hydrologic Science, Global Change and Extreme Hydrology: Testing Conventional Wisdom, Washington, D.C., 2011, p. 3.

[83] Ibid., p. 7.

[84] Ibid., p. 9.

[85] P.C.D. Milly et al., "Stationarity Is Dead: Whither Water Management?," Science, vol. 319 (February 4, 2008), p. 574.

[86] The Colorado River basin is somewhat unusual in that the Secretary of the Interior acts as water "master" for the river, and apportionment of water supplies among the basin states is done in accordance with the Colorado River Compact and a body of law known as the "Law of the River." For more information on the Law of the River, see http://www.usbr.gov/lc/region/g1000/lawofrvr.html.

[87] Tim P. Barnett and David W. Pierce, "When Will Lake Mead Go Dry?" Water Resources Research, vol. 44 (March 29, 2008), p. W03201, DOI:10.1029/2007WR006704. Note: Colorado River inflow is highly variable. Although inflow has been below average for 10 of the last 13 years, reservoir storage in the basin had increased by more than 8 million acre-feet between 2005 and 2012, reaching nearly 63% of capacity by April 3, 2012. Hydropower production continued under 2007 "interim guidelines" for managing water shortages in the Lower Colorado River basin. Reservoir management is similarly dynamic reflecting ever-changing basin hydrological conditions and predictions. See also U.S. Dept. of the Interior, Bureau of Reclamation, Colorado River Basin Water Supply and Demand Study, December 2012, http://www.usbr.gov/lc/region/programs/ crbstudy/finalreport/studyrpt.html.

[88] For more information on 2012 operations, see http://www.usbr.gov/uc/water/crsp/studies/24Month_03.pdf, accessed May 2, 2012. CRS has not determined to what degree recent scenarios are similar to those considered in studies supporting the new shortage criteria for Colorado River water allocations under the Colorado River Compact.

[89] U.S. Dept. of the Interior, Bureau of Reclamation, Colorado River Basin Water Supply and Demand Study (December 2012).

[90] For example, reauthorization of NIDIS, S. 376, and reauthorization of the Reclamation States Emergency Drought Relief Act, H.R. 518.

[91] The National Drought Policy Act of 1998, P.L. 105-199 (42 U.S.C. 5121 note).

[92] Available at http://govinfo.library.unt.edu/drought ndpcreport contents.htm.

[93] Ibid., p. 1.

[94] NOAA allocated $12.1 million for NIDIS in FY2012. For more NIDIS information, see http://www.drought.gov.

[95] Infra, note 52.

[96] Available at http://govinfo.library.unt.edu/drought ndpcreport contents.htm.

[97] Ibid., p. i.

[98] Ibid., p. vi.

[99] Ibid., p. 36.